KU-259-853

The River Cottage

Cheese & Dairy Handbook

The River Cottage Cheese & Dairy Handbook

by Steven Lamb

with an introduction by
Hugh Fearnley-Whittingstall

rivercottage.net

BLOOMSBURY PUBLISHING
LONDON • OXFORD • NEW YORK • NEW DELHI • SYDNEY

For Jean, Jean and Jeanie-Ray

BLOOMSBURY PUBLISHING
Bloomsbury Publishing Plc
50 Bedford Square, London, WC1B 3DP, UK

BLOOMSBURY, BLOOMSBURY PUBLISHING
and the Diana logo are trademarks of Bloomsbury Publishing Plc

First published in Great Britain 2018

Hook & Son (hookandson.co.uk) and Golden Cross Cheese Ltd (goldencrosscheese.co.uk) were kind enough to grant access to the photographer for this book

A catalogue record for this book is available from the British Library

ISBN: HB: 978-1-4088-7347-2

2 4 6 8 10 9 7 5 3 1

Project Editor: Janet Illsley
Designer: Will Webb
Photographer: Gavin Kingcome
Indexer: Hilary Bird

Printed and bound in Italy by Graphicom

Contents

Food is always about more than the finished dish. Every ingredient is the culmination of a journey, of a series of happenings, decisions and interventions. While that's true of pretty much everything we eat, there are certain foods that illustrate the principle particularly well, demonstrating how environment, *terroir* and technique can all alter the final result. Those foods include bread, wine, charcuterie and dairy products – especially cheese.

These are foods few of us make for ourselves any more. Simple in terms of raw materials they may be, but they also demand a certain level of skill and patience. Perhaps that is why brewing, baking, curing, smoking and cheese-making – crafts that were once practised in ordinary households – are now so far from the domestic sphere that they seem mysterious and arcane. Mostly we leave the making to distant manufacturers, to the detriment of quality and diversity. But it doesn't have to be that way. We have the option of trying our hand at these culinary arts. And if we do, the rewards can be immense, not to mention delicious.

I'm delighted to be introducing this latest River Cottage handbook on cheese and dairy. Not only is it full of fantastic and achievable recipes, it reinforces some of the key tenets on which River Cottage was established: not least that making your own ingredients is empowering, fulfilling, and a whole lot of fun. Producing some of the food you eat, even if only a little, also encourages you to think much more deeply about all the rest. A cheese-maker or bacon-curer will always be a better cheese- or bacon-shopper!

If there's one person who knows that to be true, it's Steven Lamb. Steven has been working with me at River Cottage since 2005 as a teacher, writer and events host. His broad knowledge, deep-rooted integrity and unflagging enthusiasm are legendary. And his attitude of endeavour, coupled with an almost geeky fascination with food science, made him the perfect person to write this book.

After all, dairy is not to be messed with: it's serious stuff. Milk is fragile, a little unpredictable; it demands respect. The subtleties of temperature and timing that can lead to such great differences in a finished product are not difficult to grasp – but they do require your attention. This is exactly the kind of detail that Steven finds so compelling. Once you've begun your dairying journey, with his careful guidance, you'll gain confidence – an ingredient as vital as any other in the process.

The wet and windswept islands of Britain produce some of the best milk and cream in the world. Much of this is swallowed up by the industrial dairy machine, but it's also capitalised upon by some exceptional artisan cheese-makers who produce deliciously complex cheeses. And we are blessed with small-scale dairies, often using milk from their own herds, turning out wonderful butter, yoghurt and clotted cream. Why shouldn't this great dairy tradition be expressed again in ordinary home kitchens? Making your own dairy products, at the simpler end of the spectrum at least, is surprisingly straightforward. You need very little specialist equipment and the ingredients are inexpensive and easy to find. Take the plunge into dairy, as it were, and you'll experience huge satisfaction. If your first batch of home-made labneh (see p.36) doesn't please you mightily, I'll eat my hat. And I can tell you from experience that a freshly baked pizza topped with home-made mozzarella (see p.96) is the apex of 'all-my-own-work' kitchen joy.

But the pleasures here are as much in the doing as the eating: cheese-making is – or at least should be – a craft: an enriching, creative process. This is a point that I know Steven feels particularly strongly about. He is a passionate supporter of our excellent native cheese-makers and the renaissance they have brought about in British cheese-making over the last 30 years or so. And he sees no reason why we can't all be part of it. If you like eating dairy, he asks, why not reclaim this wonderful skill from the factory farms and the mass-producers? Why not be part of the movement of bold dairy artisans encroaching on that bland, characterless territory? Making your own cheese, and choosing cheese made by passionate artisans, is the kind of food activism we can all get involved with.

How you source your raw materials makes a statement too. Milk production in dairy cows has increased outlandishly in recent decades, more than doubling in the last 40 years as the industry finds new ways to squeeze more out of each animal. This must give us pause. To take milk from an animal is always exploitative to some degree but those of us who want to enjoy dairy products have a responsibility to reduce that exploitation to the minimum. The home-dairy enthusiast is well placed to do this. If you want to make your own butter or cheese, you have to source milk of foremost quality and freshness. You're compelled to ask questions, to think about where the milk comes from and how it has been produced. As Steven explains, you have the option of choosing the very best kind of milk, from a local herd, organic or higher-welfare, perhaps even unpasteurised. You can set the bar as high as you like, and you'll never look at those ranks of cartoned milk in the supermarket in the same way again.

Few would argue with the point that butter, cream and cheese are foods we should enjoy in moderation. But I, for one, want the dairy I do eat to be the best possible. Sometimes that means buying the finest local, organic products, sometimes it means making my own and, these days, it might also mean pinching some of Steven's recipes...

So peruse these pages and see if you don't find your own interest piqued by the dairy delights within. I'm sure you will. You'll see that you can probably begin your adventures in cheese-making right away – today even – with ingredients you already have to hand. You will discover that no cheese is quite as special as your cheese. And who knows where the journey might lead you?

Hugh Fearnley-Whittingstall, East Devon, December 2017

Milk & Dairying

I've enjoyed eating dairy products for as long as I can remember. One of my earliest memories is of sitting in the garden, munching on a wedge of crumbly, pale, tangy Cheshire cheese, utterly content. Even the sound of those two words together – 'Cheshire cheese', like the noise a steam train makes leaving a station – fills me with happiness. As an adult, often the first thought I have on waking relates to dairy ingredients. When I open the fridge, particularly around breakfast time when there is a small army of children to feed, it is dairy foods that never fail to offer up a solution to hunger pangs – whether it's milk for muesli (or a strong coffee), yoghurt with berries and honey, or cheese melted over a toasted crumpet with a splash of Worcestershire sauce.

As someone who has never been able to resist the idea of having a go at things myself, it's not surprising that these days, alongside the cheese, butter and milk sourced from local producers, there are often home-made dairy products on my table – maybe creamy balls of labneh, dusted with herbs and black pepper, a hank of stretchy mozzarella, or little wedges of quick-ripening Caerphilly.

Nowadays, dairy is often the preserve of big manufacturers, but for centuries it was the business of individual households. So why not mine? And why not yours? If you feel intrigued by that question, then this book is for you. Because the truth is, anyone can make their own dairy products, whether they live in the heart of the country, surrounded by fields of grazing cows and sheep, or in a flat in the heart of the city. You just need milk, a few simple items of equipment, a little time and a bit of curiosity!

At River Cottage, we have always taken great delight in tapping into the vibrant dairy and cheese community that surrounds us. The West Country, with its mild climate and abundant pasture, has a fine dairying tradition. At River Cottage HQ you can turn 360° and, at any point, be within striking distance of a producer turning out delicious artisan dairy products from a high-welfare herd. We use many of these products, either just as they come, or as ingredients in delicious meals. This book is inspired by those people and by the craft and tradition of dairying and cheese-making that they represent.

If you take the first simple steps in dairy today, you'll soon discover that the basic principles involved in turning fresh milk into most dairy products – whether that's a very simple goat's curd or a more complex, matured hard cheese – are really very similar. The main difference lies in the time that process takes. I can assure you that making your own dairy products is a rewarding experience, and the more experience you gain, the more satisfying it becomes.

Dairying in the UK

The milk of animals has been important to us for millennia, ever since people developed the understanding that female animals could be milked by hand and dairy production began. There is reference to milking in the Hindu creation myths from as early as 5,000BC, and some evidence that simple cheese-making goes back even further than that. An archaeological dig in Poland in the 1970s discovered 7,000-year-old clay pots studded with holes which, when analysed, were found to contain milk fat molecules, suggesting that they were used for separating curds from whey.

The first yoghurts and cheeses were most likely little more than the accidental consequence of milk curdling and separating. They would probably have tasted awful. But over thousands of years, we have learnt how acids and enzymes can be used to curdle milk efficiently and contribute to its flavour; how temperature, bacterial cultures, moulds, yeasts and salt can all be used to manipulate milk into delicious and long-lasting dairy products.

For a long time, dairying was invested with a certain mystery, and attended by many superstitions. Ash-wood handles on the butter churn were supposed to foil witches' spells, while sprinkling salt in the shape of a cross at the bottom of the milk bucket would ward off evil. There was even a patron saint of butter, Saint Haseka, to whom whispered prayers could be offered in the hope that the butter would form nicely and not go rancid. None of this is any great surprise, considering the fact that there were (and are) so many variables: the very grass that fed the animals, the weather, the microclimate in the barn or the buttery or the kitchen, and the technique of the person handling the milk.

Scientific explanations of what goes on in the dairy began to emerge with the Age of Enlightenment in the mid-eighteenth century, through the research of scientists such as the German bacteriologist Ferdinand Cohn and the French chemist Louis Pasteur. These days we have a thorough understanding of what milk is made of and how it behaves in certain conditions. The guesswork and finger-crossing has been taken out of dairying, but still it has retained that sense of being a little bit miraculous. The range of products that can be created from a few litres of fresh milk really astounds me – as does the fact that they will so be locally and seasonally distinctive.

Cheese has an especially broad range – it can be fresh or matured, soft, semi-soft, semi-hard or hard and these are just umbrella terms for a whole raft of sub-categories including washed-rind, mould-ripened and blue. As human beings, we can detect five distinct tastes: sweet, sour, bitter, salty and umami (or savoury). There are not many foods that can deliver all of those tastes but cheese can, sometimes in just one slice.

Growing up in Manchester, I thought that eating cheese from Lancashire and Cheshire was something I had no choice about, just like following the local football team. We never ate Red Leicester, Wensleydale or Gloucester in our house, and certainly not cheeses from further afield, or abroad. Even though this was a rather narrow, prohibitive view of local cheese, it did give me an early understanding of cheese regionalism and a sense of how diverse British cheese production was. This diversity is one of the many things about cheese that I most appreciate now. It is one of the few British foods that has a true sense of geography and place, that can be used to map the regions of the UK.

I often feel that as a nation we have lost our identity when it comes to food, and for a while the quality of our dairy products dwindled too, their individuality swallowed up by blanket pasteurisation, industrialisation and bureaucracy. But there has been a wonderful revival in the last few decades, led by cheese pioneers such as the late Patrick Rance, and Randolph Hodgson at Neal's Yard Dairy.

Our dairy products – the best of them, that is – still represent who we are and the land we live on. They express the *terroir* of the places in which they are made in a wonderful way. A piece of cheese informs you of more than simply its integral flavour and texture. It speaks of the pasture, the people and the production. Each cheese marks a significant spot on a map. I like to imagine an invisible dairy network running up and down the British Isles, knitting it together like a curd.

The exciting thing is that if you have a fridge, a stove and a saucepan, you can tap into this regional web and produce dairy foods that express your own local *terroir*. For your first forays into yoghurt, butter or cheese-making, you may well choose standard, pasteurised milk from your nearest shop. But I hope that as you grow more confident, you'll seek out milk from local farms – perhaps even raw, unpasteurised milk which, if handled correctly, produces the most complex and distinctive dairy products of all.

I eat a lot of cheese and I love butter, yoghurt and cream. Of course, I don't make all the dairy products I consume myself. The best of these – the work of the real experts – are an inspiration for anyone interested in home-dairying. Being a small independent cheese-maker is not an easy option. It is labour-intensive and time-consuming, with the yield of product as low as around 10% of the raw material you started with. I am in awe of the passion and unerring drive to maintain traditional techniques and unique flavours that these dedicated producers exhibit. And they are winning.

Today, with the help of bodies such as the Specialist Cheesemakers' Association, there are more raw milk and independent cheese producers in Britain than there are in France – more than a thousand at the last count. It's through enjoying the fruits of their labours that my own passion for dairy, which you'll find expressed in these pages, has come about.

The ethics of dairy

Without milk, whether from cows, sheep, goats or buffalo, the culinary world would be very, very different and, I think, far less rich and interesting. But those of us who enjoy milk and its products, and who want to continue consuming them, cannot shy away from the ethics and welfare issues that go with them.

Human beings have 'progressed' from milking small herds, or even single beasts – perhaps sharing that milk with the calf, lamb or kid still at its mother's side – to taking vast quantities of milk from hundreds of thousands of animals on an industrial scale. Milk these days is abundant and cheap. Its very cheapness and ubiquity mean that an ingredient that once held high status now hardly registers when we put it into our shopping basket. Milk has been industrialised and because of this, we have stopped thinking of it as the living product of a living animal, so the thorny issues surrounding milk and its production are often passed over.

Milk is produced only when a female animal gives birth. In order for milk to be available to us on such a huge scale, our dairy industry relies on the impregnation (mostly through artificial insemination) of hundreds of thousands of cows every year. These cows' calves are removed from them very shortly after birth in order that we may benefit from the milk intended for those offspring. The cows are milked two or three times a day for the following 10 months until, after a rest period of a couple of months, they are impregnated again. This continues for between 4 and 13 years for each cow until its yield reduces to the point where it is no longer profitable to feed it, after which it is culled and sold as cheap beef.

There's no getting away from the fact that all dairy production is exploitative to some degree. But as a consumer of milk, you don't have to support poor animal welfare and a volume-over-quality approach. There is a wide spectrum in dairy farming in terms of intensiveness. The worst examples are the farms that have no consideration for the wellbeing of their animals, which are merely a commodity to be profited from. Very intensively farmed cattle are bred to be high-yielding, have little or no access to pasture and are fed entirely on proprietary feeds, including 'concentrates' rich in calories and protein, to boost the production of milk.

Cows from 'high-output' systems like these are likely to produce much more milk than cows on organic dairy farms, but they are also more likely to suffer health problems. Cows on a non-natural diet are more likely to develop acidosis, a digestive issue, and the preference for grain feeds has also led to an increase in the number of instances of E-coli and salmonella found in the stomachs of cows. Animals kept indoors for long periods may have weakened immune systems, and diseases are more easily spread. To counteract problems such as mastitis, intensively farmed animals are regularly dosed with antibiotics, which then pass into the human food chain and have been linked to antibiotic resistance in humans.

The best dairies are able to maintain high animal welfare while still making a commercial success of the operation. Soil Association certified organic cattle, for instance, must be allowed to graze outdoors for at least 60% of their time and the routine use of antibiotics is banned.

The dairy goat industry exists on a much smaller scale but is not without its faults. The majority of goat's milk comes from zero-grazing, indoor herds; most dairy goats never go outside. This is because goats are simply not as well suited to the British climate and environment as cows and sheep and it is difficult to make a herd productive and profitable if it is outdoors. For the consumer, the best solution to this issue is to buy organic; by doing so, you are at least supporting the highest welfare standards.

Every type of dairy industry has the difficulty of excess males. Almost every male calf or kid born to a dairy mother is surplus to requirement because they will not go on to produce milk themselves. The traditional solution has been to bump these males on the head shortly after birth and dispose of the carcasses. There's a lot being done to tackle that shocking waste, with more male calves now being raised for beef and a nascent kid-meat industry developing too. But the conundrum is far from solved.

When you consider the enormity of issues like these, it is easy to feel powerless. But once you remember the fact that both problems and solutions come from people, not merely faceless systems, it becomes less hopeless. So, search out products that are being made by people with similar ideals to your own, sold in places where high standards are achieved. I would suggest choosing either organic milk or milk that is produced locally, on a small scale by people who handle and process it themselves, as opposed to having it collected, transported and amalgamated with the produce of many other dairies. The best milk is produced from pastured animals, and is minimally processed. This is also the milk most suited to making cheese and other dairy products.

Different kinds of milk

Milk is a complex liquid, comprised of a variety of minerals, proteins, sugars and fats. The production of milk is one of the main things that distinguishes mammals from reptiles; it allows the more or less helpless offspring of mammals to continue their development after birth with the benefit of an on-tap, complete source of nutrition (whereas reptiles are born fully formed and ready to go with little reliance on their mother).

Although all mammals produce milk, only a select few are used for dairy purposes. The cow is the most common, followed by sheep and goats on a fairly equal footing, and then buffalo. All these animals are ruminants with long, prehensile tongues that can grasp blades of grass, and have the ability to feed on relatively poor and dry pasture and still produce copious amounts of milk.

Once a cow (or other mammal) has given birth, the mammary glands take nutrients, proteins, sugars and fat from the blood pumping around the mother's body to make milk. This is drained through thousands of filters towards the teat.

The very first secretion of fluid through the udder is a thick, creamy substance called colostrum. This is vital for the new-born, because it is packed with protein, vitamins and fat, as well as antibodies to ward off infection. In dairy farming, the calf is removed from the mother once this colostrum has run out, about 3 days after birth. The colostrum will have already depleted significantly after the first 24 hours. What comes next is the saleable product, milk.

Whether from cattle, sheep, goats or buffalo, milk always contains at least 80% water. Beyond that, the quantity and quality of fat, lactose and protein in different animal milks differ considerably. Milk from all these species can make great dairy products, though some are specifically chosen for certain products, such as buffalo milk for mozzarella or cow's milk for Cheddar. The considerations range from availability, freshness, fat content and colour to nutritional value. Differing amounts of milk solids (principally fats and proteins) are a major factor; this affects the yield of cheese. Sheep's milk, with the highest percentage of solids at about 20%, guarantees the greatest return in terms of yield.

The quality of milk is dependent on what the animal eats. The best milk, which will go on to make the best dairy products, will always come from animals that have been fed their natural diet as part of a high-welfare system that encourages natural behaviour. Those pastured on fresh green grass, clover, foliage and herbs may not produce higher yields of milk, but it will be better quality than milk from animals given grain feed or silage and kept confined.

Goats aren't particularly picky eaters but they do like the taller weeds which cows don't, and in terms of pasture management using goats as a 'follow on' herd will benefit the pasture. As long as goats are given a bit of variety to browse on, they

will do well and most pasture has enough diversity to entertain this. Fresh green grass is the best source of minerals and vitamins for all dairy animals – vitamins and minerals are important because they help the curd to develop when you're making cheese. It would be impossible, for example, to make good Alpine-style cheeses with milk from a cow that has never seen grass.

There is also clear evidence that milk from pasture-fed or organic animals is higher in certain nutrients than conventional milk. These include vitamin E, beneficial omega-3 fats and conjugated linoleic acid (CLA) – good fat which may help protect against some serious health conditions such as diabetes.

Cow's milk

Most of the dairy products we are familiar with in the UK are made from cow's milk – it is good for butter, yoghurt and cheese of all kinds. The milk that cows produce has a grassy, sweet character, and they make more of it than any other animal – somewhere between 50 and 60 litres a day. However, it has a very high water content and much of this is lost during the cheese-making process so the yield is relatively low – the average yield for cow's milk cheese is 10% of the weight of liquid milk used.

The most popular dairy cattle breeds in the British Isles are the Friesian and the particularly high-yielding Holstein (or a cross of these two breeds); these are the classic, black-and-white-patched cows that are so familiar to us. Jerseys and Guernseys are renowned for the creaminess of the milk, while in Scotland the Ayrshire cow's milk is preferred for making cheese.

Goat's milk

Goat's milk has a distinctive smell and flavour. Some people prefer to drink it instead of cow's milk because it is closer to human milk and the fat globules are smaller, which aids digestion. Digestibility is also helped by the high amount of medium-chain fatty acids in goat's milk. Goat's milk contains less lactose (milk sugars) than cow's milk, which is preferable for those who suffer from lactose intolerance. Unlike cow's milk, which is slightly acidic, goat's milk is slightly alkaline (this does appear to lower the risk of allergies or digestive issues caused by inflammation).

On average, a healthy goat produces 3–4 litres of milk a day. Goat's milk is very rich and exceptionally high in nutrients – despite the fact that, these days, it will almost certainly be from a zero-grazing animal. If pastured, goats can produce milk that rivals cow's milk when it comes to forming good curds during the cheese-making process. However, forming strong curds is not the usual quality of goat's milk and a lot of goat's cheese recipes include the addition of calcium chloride to help strengthen the curds.

Cheese made from goat's milk is often referred to as *chèvre*, from the French for 'goat'; this is the cheese of choice throughout most of central, rural France. Goat's cheeses have a characteristic minerally taste that ranges from mild to strong. They can be described as sour or tangy, and perhaps even a bit spicy. Goat's cheeses tend to have a paler, whiter flesh than cow's, which is due in part to the lower levels of carotene in the milk.

Sheep's milk

Sheep produce creamy-yellow milk that has a slight shimmer due to the high amount of protein and fat in it. The fat in sheep's milk is dispersed through the liquid (as opposed to forming a thickened layer on top) and it is easier to digest than cow's milk.

Sheep's cheeses are usually rich, a little oily and slightly golden in colour. The flavour is often directly of the farm, and nutty, with scents of dry grass or sweet silage. While sheep produce only around 2 litres of milk per day, when you are turning it into dairy products, the yield is high, owing to the impressive fat, protein and sugar content. The substantial levels of butterfat make it particularly good for producing blue cheeses, such as Roquefort or Beenleigh Blue.

Buffalo milk

Water buffalo produce around 16 litres of milk a day, which is surprisingly low when you consider their size. There are no examples of intensively farmed buffalo in the UK – the largest herd, at Laverstoke Park Farm in Hampshire, consists of around 300 animals. Free-range buffalo are not prone to disease and the quality of their milk is generally high. Buffalo milk has 50% more protein than cow's milk, 40% more calories, nearly 40% more calcium and high levels of the natural antioxidant tocopherol, which, manifested as vitamin E, can promote healthy skin. It also has twice as much fat – an even spread of saturated, monounsaturated and polyunsaturated. The fresh milk requires some degree of homogenisation (see p.28) or the fat will solidify on top of the milk within a couple of days.

The high fat content of buffalo milk makes it particularly good for cheese- or ice-cream-making. The milk is sweet, floral and aromatic and has perhaps the cleanest flavour of any of the farmed milks. These flavour characteristics are reflected in the cheeses it produces.

Pasteurised milk

Pasteurised milk, which is by far the most common milk on the market, has been heat-treated to kill off any pathogenic bacteria and spoilage microbes, making it more safe and stable and extending its shelf-life. There are different categories of pasteurisation, depending on the duration and temperature of heating.

- Ultra-high temperature (UHT) milk is heated to 130–150°C in a matter of seconds – which means it can be stored outside of refrigeration for months. This is the milk that sits on the non-refrigerated supermarket shelves or is served in small, impenetrable plastic pots on trains.

- Sterilised milk is zapped at around 120°C for half an hour and is devoid of any microbe activity. This long-life milk can be stored, unopened, at room temperature indefinitely and is less white in colour than fresh milk, with a strong 'cooked milk' flavour. It is, to the dairy enthusiast, dead milk.

- Standard pasteurised milk has either been heat-treated to 72°C for 15 seconds or, less commonly, to 63°C for 30 minutes. This gets rid off any pathogens but the milk is not sterile and some bacteria remain, which means it will eventually go off. The main concern for the cheese-maker who wants to use pasteurised milk is freshness. Most milk on the supermarket shelf is already at least 3 days old. Pasteurisation may also impair or slow down the milk's ability to coagulate.

Some small dairies are now pasteurising their own milk on the farm and, hopefully, as support for local producers increases, this trend will continue. It is agitated far less than milk that has been pumped, piped and transported from farm to depot to bottling plant – a factor many traditional cheese-makers believe has a huge impact on quality. It can also be sold much fresher and, to my mind, this is a great option for making dairy products. Search out small, independent farm dairies online. The quality of this milk should be fairly consistent but you might notice that cheese made with it does vary from batch to batch, even if you are following the same method. This is something to be celebrated and embraced.

It is possible to pasteurise raw milk at home if you are looking to become more self-sufficient. To avoid scalding the milk, it's best to place the pan of raw milk inside a larger pan containing 6–8cm water, then maintain it at a constant temperature of 63°C for half an hour. However, it can be hard to achieve this with a domestic cooker. I commend any attempt to gain as much control over the dairy process as possible, but realistically I think that it's better to use either good raw milk that you can trust (see p.24) or to source farm-pasteurised milk.

Raw milk

Unpasteurised or 'raw' milk has not been treated, except for being cooled to 4°C. It is packed full of bacteria and microbes, which lead to a whole host of interesting dairy flavours – and, in cheese, textures as well – but it can also present a serious health threat if not handled in the correct way. Raw milk, even to the enthusiast, shouldn't be seen as the Holy Grail of milk just because it is raw. More than any milk, it needs to be scrutinised for quality. Also, it's definitely not suitable for anyone with a compromised immune system, or for pregnant women, the very young or the elderly.

Unpasteurised milk was more or less normal in Britain – though decreasingly so in the cities – until the 1950s. Nowadays, it is fairly hard to find. The sale of raw drinking milk (RDM) is prohibited altogether in Scotland and there are strict regulations around it in England, Northern Ireland and Wales, where it can only be sold directly by the producer, and only if the dairy has the correct licence. The container needs to display an 'appropriate' health warning.

Raw milk must be handled with care, in a hygienic environment. It can harbour harmful bacteria, such as listeria, E-coli and salmonella, and can even cause severe illness, although this is rare. Registered farms who are licensed to sell raw milk are under extra scrutiny, which means they do have to check very regularly for potential pathogens.

Some would argue that raw milk is intrinsically 'healthier' than pasteurised, because none of its natural enzymes and nutrients have been damaged by heating. Whether you agree or not (and I don't think there's clear evidence either way), raw milk delivers a depth and complexity of flavour surpassing any other milk, whether drunk or turned into cheese, and retains unique qualities reflecting regionalism and seasonality that would be 'ironed out' by pasteurisation.

Raw milk cheeses

Pasteurisation will always affect flavour and microbe diversity in milk. This is particularly relevant in cheese-making. There are some cracking cheeses made from pasteurised milk (Colston Bassett Blue Stilton being one of my favourites), but raw milk cheeses celebrate the artisan nature of cheese and promise real depth of flavour and regional diversity.

Raw milk remains the very soul and spirit of traditional cheese production and there is a strong resurgence of unpasteurised cheese-making in the UK. Not everyone is comfortable with this – how much of a risk it might pose is often put to governmental review – but industry regulations are strict. Professional raw milk cheese production undergoes microbiological monitoring and is regulated and checked more than regular cheese production.

Proponents of raw milk cheeses, such as Randolph Hodgson of Neal's Yard Dairy and natural-cheese-maker David Asher, insist raw milk produces better and more interesting cheese. I agree.

It is science that has safeguarded the production of cheese, with discoveries such as pasteurisation, and it was the creation of starter cultures in laboratories that enabled producers to precisely control levels of acidity so that bad bacteria could not thrive. However, the microbial communities in raw milk, of which very little was known until more recently, were depleted in the process. This contributed to a restriction in the range of cheeses available in the UK that shackled the industry, despite good intentions.

While we know how to kill microbes, we are only now beginning to understand the possible value of keeping them alive. Recently, for instance, it has come to light that there are strains of microbes naturally present in some milk that are resistant to bad bacteria. This could mean that there is a whole new world of delicious raw milk cheeses waiting to be discovered, which would be reliably safe to eat.

There are already some excellent raw milk cheeses, such as Salers, which is produced in the French Alps. Salers is made from raw milk left to ripen in wooden vats without any additional starter cultures and it has been shown to contain microbes that can see off all the bad bacteria, including listeria. The wooden vats are never cleaned: the communities of bacteria in the milk and the wooden barrels work together, acidifying the milk to create a complex, unique, delicious and safe artisan cheese.

Milk fat

The fat in milk takes the form of microscopic globules, each surrounded by a protein membrane. The fat globules are less dense than the rest of the milk and they rise to the top and settle in a layer (except for sheep's milk, where the fat is evenly distributed throughout). This effect, called 'creaming', is the reason birds used to peck away at the foil tops of milk bottles in order to get at the fat-rich layer underneath. The skimming off of this naturally separated fat is the first stage in producing cream or butter.

The amount of fat largely dictates how many calories a glass of milk will contain and how creamy it will be. The fat content also provides the framework for dairy products: the fattier and creamier the milk, the more suited it is to making cream, butter and cheese.

The pairing of fat and protein in milk benefits the cook, in that it affords the milk a good deal of tolerance when it is heated. The protein membrane around the fat globules stops them from bursting, which would release fat into the milk and would make it prone to spoiling. As the heat increases, the proteins form stronger, thicker bonds around the fat, which is why whole milk is perfect for creamy sauces, custards and reductions.

The protein casing on the fat globules can be damaged, however, by agitation – such as the stirring, pumping and transportation of milk that's being produced on an industrial scale. These processes can break open the fat globules, causing lipolysis (the breakdown of fats), which may affect the flavour of the milk. Keeping milk as 'whole' and undamaged as possible makes a difference to the quality of the products made from it.

Fat gives whole milk a luxurious texture, more flavour than skimmed milk, and more nutrients. Milk fat contains the fat-soluble vitamins A, D, E and K. Because these are non-soluble in water, they are retained in reserve in the body without necessarily having to be frequently topped up. They aid normal organ function and the repair of body cells. Vitamins A and D are particularly potent, as they both work on bone growth by aiding the absorption of calcium. Vitamins E and K help to maintain healthy blood and support the immune system.

Homogenised milk

Homogenised milk has been treated to prevent the natural 'creaming' process where a layer of fatty cream rises to the top. In order to distribute the fat evenly throughout the milk, it is heated and then fired at high pressure through a pipe with several small nozzles. This process causes the fat globules to split into smaller molecules, lose their protective membrane and become attracted to the casein proteins that milk also contains, and which will weigh them down. The casein-weighted fat globules will now be distributed more evenly through the liquid.

All homogenised milk is also pasteurised (see p.23) because the rupture of the membranes around the fat globules makes them vulnerable to attack from enzymes that can cause rancid flavours. Pasteurisation kills off these microbes.

Milk proteins

There are many proteins in milk and they can be classified into two main categories: casein or whey. There are four times as many casein proteins as whey proteins, and it is the casein proteins that enable milk to transform from a liquid into a thickened, viscose yoghurt or a solid mass of cheese. The casein proteins in milk form groups or families called 'micelles' which are held together by calcium. These account for a tenth of the volume of cow's milk.

Pints
fl.oz
ml
2
40
3/4
35
1000
1/2
30
900
1/4
25
800
1
20
700
600
500

The micelles start to alter when acidity and temperature are increased. Cow's milk is already a slightly acidic liquid at pH 6.5 but increasing the acidity – to around pH 5.5 – makes the micelles join together to form loose communities. However, at this point, the calcium glue that helped form the micelles in the first place starts to dissolve. This exposes the individual casein proteins and they are dissipated. As the acidity increases (i.e. the pH lowers) towards pH 4.7, and the temperature climbs, the casein proteins start to bind together again, not in loosely connected families this time, but as a unified crowd, producing solid 'curds'. These are the basis for most cheeses.

Whey proteins, meanwhile, do not respond to heat like casein proteins. They are left suspended in the remaining liquid, which is itself known as 'whey'. Some cheeses are made from whey proteins, including ricotta (see p.102).

A2 milk

There is currently some debate around the two forms of beta-casein protein that are contained in milk from different breeds of cow. 'A1' beta casein, which is present in the milk from the most common dairy breeds (including Holstein and Friesian, which dominate our market), produces a peptide in the human gut, which some believe may have negative effects on immunity and digestion. This is particularly relevant for those who have low lactose tolerance.

'A2' beta-casein is the more common protein in milk from cattle breeds that originate from the Channel Isles and Southern France, such as Jersey, Charolais and Limousine. 'A2' milk is sold as a healthier alternative to standard milk in many supermarkets nowadays. However, scientific research into this issue is not, as yet, conclusive.

Cow's milk allergy

This is the most common form of food allergy suffered by infants and young children; most grow out of it as they get older, generally before they start full-time school. It is triggered by the protein in cow's milk and should not be confused with lactose intolerance (which is an adverse reaction to the sugar in milk, see opposite page). Unlike lactose intolerance, cow's milk allergy involves the immune system and common symptoms include skin reactions, nausea, vomiting, colic, a cough, runny nose and wheezing.

The general medical advice is to avoid withholding dairy products unless there is an adverse reaction (i.e. do always try children on cow's milk to start with, even if there are allergies in the family). No dairy products should be given to children under the age of 6 months, however, and they should only be introduced after that as a component of meals. Milk as a stand-alone drink can be given to infants over the age of 12 months.

Lactose

Milk's natural sweetness comes from lactose, which is a combination of the 'simple' sugars glucose and galactose. Lactose is unique to milk, forged in the secretory cells of the mammary glands. It is the conversion of lactose to lactic acid, via fermentation, that begins the process for making most dairy products.

The longer milk is left to acidify (or sour) at the beginning of yoghurt- or cheese-making, the less lactose it contains. Most of the remaining lactose is contained in the whey, not the curds. Softer, moister cheeses have more lactose; harder, matured cheeses have little or none.

Lactose intolerance

Lactose can be problematic: it can only be digested by humans if good levels of the enzyme lactase are present in the gut. The vast majority of the world's population, particularly people of African and Asian descent, only produce low levels of lactase beyond early childhood (i.e. beyond the age of weaning off the mother's milk) so they have a reduced ability to digest lactose and are therefore lactose intolerant. Those who can continue to digest milk into adult life do so because of genetic adaptation. The most common symptoms of lactose intolerance are abdominal discomfort, bloating and cramps. It is not an allergy and it is not life-threatening.

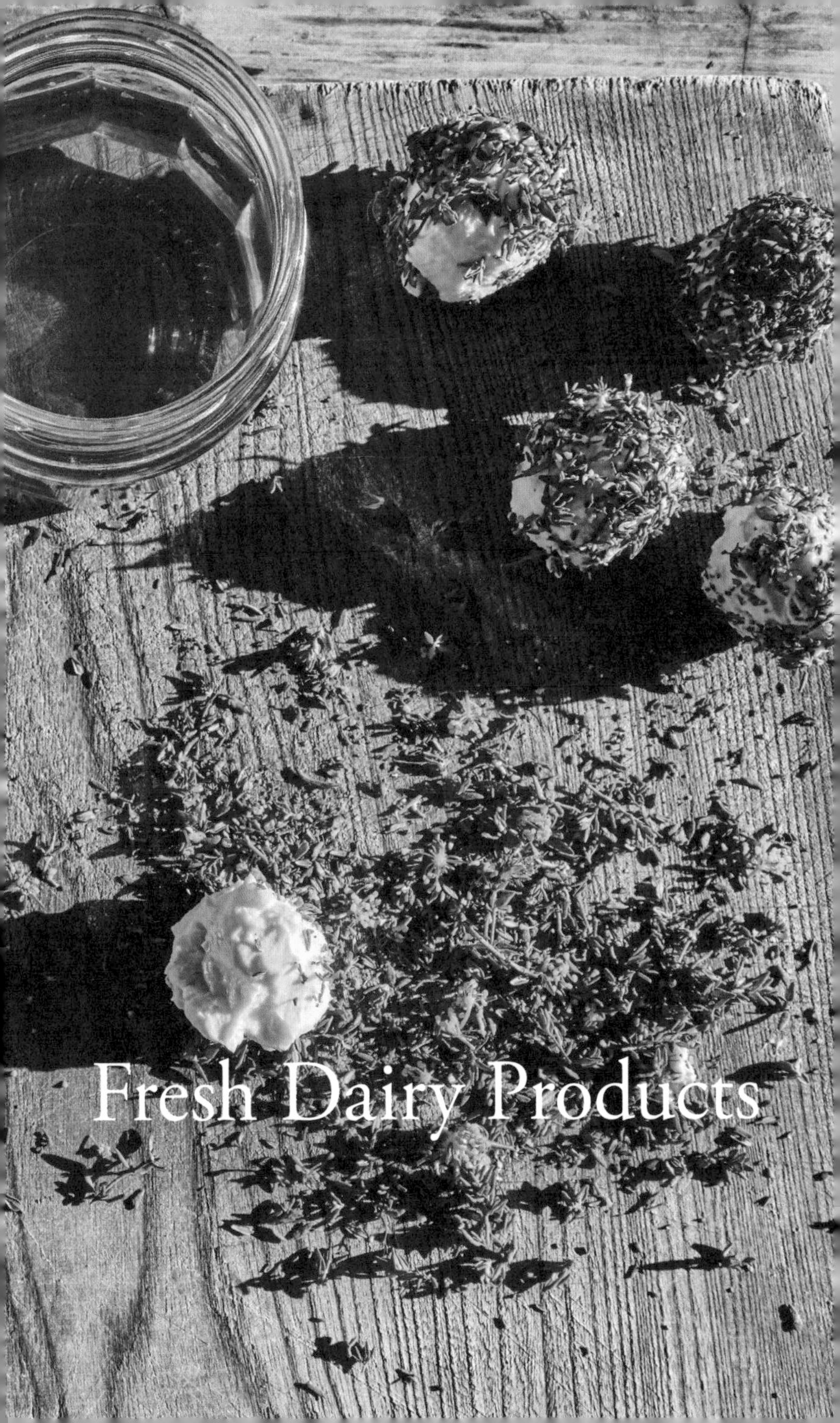

Fresh Dairy Products

Butter
and buttermilk

This is a simple recipe for creamy butter that requires only one or two ingredients and a bit of elbow grease. The process involves beating double cream until its fat globules start to break down. The pure fat masses together into butter, and leaves a milky liquid – buttermilk – behind. The buttermilk can be stored in the fridge and used for making the scones on p.197.

Makes about 250g
500ml double cream
1 tsp good-quality sea salt

Pour the cream into a bowl and leave to stand until it loses its fridge coldness.

Now start whisking the cream, using an electric whisk. Once you've passed the stage of a stiff whipped cream, lumps will begin to appear in the mix. These will start to separate from a milky liquid, which is buttermilk. Drain this off and keep it in the fridge for up to 7 days (to use in other recipes).

Pour enough cold water over the lumps left in the bowl to just cover them. Return to whisking vigorously so that the lumps start to bind together. The cloudy, watery liquid that separates out can again be drained off. Discard this, as it is mostly water, rather than buttermilk.

Add a little more cold water to the mix, whisk again, and drain. Repeat until the liquid runs clear, which indicates that the butter is ready for kneading and compacting. Depending on how vigorous you are when whisking, this should take 5 minutes or so. The butter should have developed a lovely yellow colour. Pour off all the excess clear liquid, then season with the salt.

Knead the butter with your hands until it becomes a single mass. You can leave it like this, or pat or shape it with a couple of spatulas. You can enjoy it straight away or keep it in the fridge, wrapped in parchment or cling film, for 3–4 days.

P.S. The flavour of this fresh butter is superb but it doesn't have the keeping qualities of a commercial butter because it retains more moisture and residual lactose. The moisture facilitates microbial action, which leads to the breakdown of fatty acids and produces 'off' and cheesy flavours. So much of dairying is about controlling spoilage – too much control results in a very safe but bland product, while less stringent control can enhance flavours but reduce stability and shelf-life.

Labneh

Simple to make, labneh is a strained yoghurt cheese with a spreadable texture and a sour, yoghurty flavour. All that's required for this recipe is yoghurt and salt – which encourages the whey to drain off – plus a large square of cheesecloth or muslin and a little patience. The only difference between labneh and Greek yoghurt (see p.43) is the addition of salt. I make labneh all year round but it is a particular family favourite at Christmas; I prepare it on Christmas Eve ready to serve the next morning with home-cured, smoked wild salmon and warm bagels. It is also delicious in salads or with ripe figs, or simply drizzled with olive or rapeseed oil.

Makes about 200g
300ml natural yoghurt (full- or low-fat)
A generous pinch of sea salt

Lay a sheet of muslin or cheesecloth over a sieve and suspend it over a bowl. Pour the yoghurt into the cloth-lined sieve and stir in the salt. The whey will begin to leak out almost immediately.

(I sometimes tie the cloth in a knot and hang it from the tap over my kitchen sink with the bowl underneath. Every once in a while, I give the tied cloth a little squeeze to accelerate the draining of the whey, but this isn't necessary unless you particularly enjoy the tactile element of it, as I do.)

Allow the yoghurt to drain overnight, either in the fridge or over the sink. The next day, you'll have a lovely ball of creamy white cheese in the cloth. You can eat it immediately or keep it in the fridge in a sealed container, where it will continue to lose whey and thicken up more, for up to 5 days.

P.S. To give the labneh a whole new range of flavours, try stirring in or sprinkling on herbs, flowers, garlic or seasoning. Chive flowers, cracked black pepper and finely grated garlic (use a Microplane) work particularly well.

P.P.S. You can preserve the labneh by rolling it into small balls and immersing them immediately in good quality oil in a sterilised jar (see p.40) with a screw-top lid. The labneh will keep for up to a couple of months in the fridge, but you'll most likely eat it within a week or two.

Clotted cream

I prefer to avoid the discussion about whether clotted cream goes on the scone before the jam or after, and instead just revel in its deliciousness. Clotting and crusting takes place when rich double cream is gently heated – it's barely even a recipe. But, simple as it is, the resulting product is wonderful, creamily complex and sweet. Shop-bought double cream works perfectly well but if you can get hold of fresh, unhomogenised Jersey milk, skim the cream off and use that.

Makes about 175ml

350ml double cream or Jersey cream

Preheat the oven to 80°C/Gas mark ¼ (or 90°C if that is as low as it can go).

Pour the cream into a shallow, wide ovenproof dish. Put the dish into the oven and allow the cream to bake very gently for about 3 hours. A skin will form, which will turn golden-yellow and eventually set. You can 'wobble' the dish to check it has set.

Remove from the oven, allow the cream to cool slowly and then transfer to the fridge. It will keep for 4 days or so – but is likely to magically disappear before then! Use to adorn scones in whichever way your regionalism dictates.

Crème fraîche
and mascarpone

Crème fraîche is cream that has been treated with a starter culture, which thickens it and produces a delicately sour flavour. (In fact, it is on the way to being a cheese, the key difference being that it is neither 'renneted' nor drained). You can enjoy the thick texture and lovely flavour of crème fraîche on the side of puddings, or use it in cooking without fear of it splitting, unlike standard cream.

If crème fraîche is drained overnight, it forms mascarpone, a fresh cheese that is delicious in a tiramisu (see p.202).

Makes about 500ml crème fraîche or 250ml mascarpone

500ml double cream
1 unit of heterofermentative mesophilic starter Flora Danica (Hansen) or equivalent (see p.70)

Sterilise a 500ml jar (see below), then stand the jar in a heatproof bowl. Heat a medium saucepan of water to 25°C (use a thermometer to check), then pour it into the bowl so that it comes a third of the way up the sides of the jar.

Pour the cream into the jar, then add the starter, stirring it in gently before sealing the jar with the lid. Leave the jar of cream in the water and allow it to cool down slowly at room temperature. It will thicken up gradually.

After about 10 hours, the cream will have thickened considerably. The pH will also have lowered – you can test this with a pH meter if you have one. It should register less than 6, ideally around 4.7. When this stage is reached, tip away the water in the bowl and replace it with ice-cold water to rapidly cool the cream. The crème fraîche is now ready to use and can be kept refrigerated for 7–10 days.

For mascarpone Spoon the crème fraîche into a cheesecloth- or muslin-lined sieve and set it over a bowl (as for labneh, see p.36). Place in the fridge and leave to drain overnight. The following day you will have a smooth lump of mascarpone, which will keep for 7–10 days in the fridge.

To sterilise jars Immerse them (and their lids) in a pan of boiling water and bring to the boil; or wash in hot, soapy water, rinse well and dry in a low oven; or put them through a hot dishwasher cycle. (For Kilner jars, remove the rubber seal first.)

Yoghurt

A simple, stirred yoghurt is a great way to begin making your own dairy products. A starter culture is added to warmed milk to acidify and set it – the trick is to keep it at a constant temperature once the starter has been added. It is possible to use raw milk (see p.24). Once you have made your first yoghurt, you can inoculate each new batch of milk with yoghurt from the last time instead of using a starter. The method is described overleaf.

Makes about 500g

500ml whole cow's or goat's milk (unhomogenised)

1 unit of thermophilic yoghurt starter (see p.70) or 30ml active kefir starter (see p.72)

Pour the milk into a medium saucepan and heat gently, stirring constantly with a spoon, until it reaches 85°C. Take regular readings in several places around the pan, rather than leaving the thermometer suspended in one place. Remove from the heat, put a lid on the pan and leave for 5 minutes.

Now stand the pan in a large bowl filled with cold water to bring the temperature down. Leave until it has cooled to 43°C.

At this point, stir in the starter. Now either place the pan over a very low heat or in a warming oven, or pour the milk into a pre-warmed flask to maintain a constant temperature of 40–43°C. Acidification should take place within 3–4 hours so wait until the milk shows signs of thickening, or until it registers a pH of 4.7, then pour into a sterilised jar (see p.40) and close the lid.

Leave the jar at room temperature until the pH has further lowered to 4.6, which should take less than 10 minutes. The yoghurt should now be chilled to stall any further acidification.

Once chilled you can eat it, or store it in the fridge for up to 2 weeks. You can also store the yoghurt in small individual sterilised jars – pour a layer of fruit purée into each jar before adding the yoghurt, if you like.

For Greek-style yoghurt For a thicker texture, drain the yoghurt in a muslin- or cheesecloth-lined sieve, as you would for labneh (see p.36).

Extra easy yoghurt

This very simple way of making yoghurt involves adding a little bit of pre-existing live yoghurt to heated and then cooled milk. Using a dollop of home-made yoghurt to start with is an appealing idea – like a yoghurt family tree.

Makes about 500g

500ml whole cow's or goat's milk (unhomogenised)
1 tbsp natural yoghurt

Pour the milk into a large saucepan and warm slowly, stirring continuously with a spoon, until it reaches 85°C; this should take around 20 minutes. The longer the milk is heated and stirred the thicker the yoghurt will be, so it's worth investing some time. Once the temperature is reached, remove the pan from the heat and sit it in a bowl of cold water to lower the temperature to about 40°C.

Add the yoghurt to a sterilised Kilner jar (see p.40) or flask, pour in the warm milk and stir. If using a Kilner, close the lid and place in a warming oven or, if using a flask, put on the lid and leave at room temperature. The acidification of the yoghurt takes about 4 hours.

The yoghurt is ready when it is set and registers a pH of 4.6; you can refrigerate it when it reaches this stage if you like, which will halt the acidification. The longer you leave the yoghurt to acidify the more tangy flavours it will develop. It will keep in the fridge for about 2 weeks.

For Greek-style yoghurt Simply drain the yoghurt in a muslin- or cheesecloth-lined sieve, as you would for labneh (see p.36) to make a thicker yoghurt (shown right).

For herb-flavoured yoghurt Put 100g Greek-style yoghurt into a food processor with 1 finely grated garlic clove (use a Microplane), 15g each finely chopped mint and coriander leaves, and 1 tbsp olive oil. Blitz until smooth and green, transfer to a serving bowl, cover and chill until required. This is delicious as a dip but works equally well on the side of spicy dishes, as it is clean and creamily soothing.

Making Cheese

There are a number of variable factors that can affect the way in which cheese develops from the milky liquid through to a ripe, semi-solid state. Not all of them are positive or even intended. Some of the first cheeses I made were closer to failure than success but then others showed certain qualities that came about through happy accident. There are a set of guidelines (rather than hard rules) that enable you to make cheese, and I have come to accept that each cheese I make will not be exactly the same as the last one but instead will have a character all of its own. Key aspects of cheese-making are covered in the following pages, and I have also included a troubleshooting guide on pp.208–13, which should help you in those moments when the dairy adventure has gone awry.

Cheese and the seasons

With unlimited, pasteurised, homogenised milk now being the norm all year round, it's easy to forget that milk naturally undergoes seasonal changes. These are only really detectable in milk from independent dairies where the animals are allowed to graze outdoors most of the year. It's the changes in what the dairy herd is eating in the pasture that makes a difference.

Spring brings an abundance of growth and grass full of nutrients. Milk will have increased fatty acid and CLA (conjugated linoleic acid) levels at this time of the year. Milk yields aren't necessarily high, however, so spring is the time to consider making small, fast-ripening cheeses.

Summer means pasture will be at its established best and will bring the highest yield of milk. This is the time to be bold and make larger, harder cheeses, primarily because milk is abundant. Traditionally, making hard cheeses during the summer is a preparation for the less bountiful winter months.

Autumn is the perfect time for making mould-ripened blue and microbial rennet cheeses. It is also the time to try wrapping a cheese in leaves or bark, which protects the cheese and helps it to retain a constant level of humidity. Such wrappings also impart earthy, nutty flavours.

Winter brings the end of the pasture-fed milk run and milk production will drop dramatically. At this time, I return to making small, fresh cheeses, dressing them with smoked garlic, black pepper or anything with a kick or a bit of heat. My cheese 'cave' will be full of ripening hard cheese at this stage, and that is where I head to with thoughts of hot, melty cheese, such as raclette.

Testing renneted milk in water

Beginning a cheese

The majority of cheese recipes start with warming milk or cream to a specified point – around 32°C for most, though it can be lower or higher. Because the temperature directly affects the action of starter cultures and rennet – and therefore the final outcome of the recipe – you should always take a gentle approach to warming milk. This makes it easier to reach and maintain the right temperature, rather than overshooting the runway.

Adding the starter culture

Once the correct temperature has been reached, a starter culture is usually added. As the natural sugars contained in the milk (lactose) are converted into lactic acid by the friendly bacteria in the starter culture, the milk becomes more acidic. This fermentation period (culturing of milk) is extremely important – the rising acidity helps to prevent harmful bacteria (pathogens) developing.

A pH meter tells you when a safe level of acidity has been reached. You are looking for a pH of 4.6 or lower (the lower the pH, the higher the acidity). The level of acidity also sets the course for the taste of the final product. In my cheese recipes I've given an idea of when to expect certain pH readings where appropriate.

Renneting

Once there has been a marked drop in the pH (usually about an hour after the starter culture is added), rennet is stirred in to speed up the process of coagulation – when the milk separates into solid 'curds' and watery 'whey'. The general rule of thumb is to add 10 drops of liquid rennet per litre of milk but some recipes give slightly different amounts for specific texture and flavour outcomes. The coagulation process can take anywhere between 15 minutes and 24 hours, depending on the temperature and freshness of the milk and the strength or type of rennet.

Flocculation

This is the stage at which the curds start to set after the rennet has been added. It can be assessed by placing a light plastic dish on top of the renneted milk, where it will float around on the surface. When it stops moving, flocculation has been reached. Alternatively, dip a finger in the renneted milk then let the drops fall off your finger into a glass of warm water. If they form little globules or flakes that sink (see opposite), this indicates that flocculation has been reached. At flocculation you will also notice a more pronounced lowering of the pH on your meter.

The curds will continue to firm up after flocculation. To test when they have set sufficiently to be cut, do the 'split test': insert a knife into the curds and turn the blade 45°: if it makes a clean break your curds are ready for cutting (see overleaf).

Cutting the curds

1

2

The set curds must be carefully cut into pieces (still in the original container), to enable the whey to drain off. Begin by making an incision in the surface of the curds, with the knife angled at 45° (pic 1). Turn your knife (pic 2) and push it back along the cut (pic 3); it should split cleanly.

Now make parallel cuts through the curds, down to the bottom, spacing them equally. Then cut across these lines to create a grid pattern (pic 4); in effect this divides the curds into vertical columns.

Next you need to cut each column of curds into small sections from the top of the pan down to the bottom. This is a bit tricky because you are doing the majority of the cutting 'blind', but with some practice you'll get the hang of it. I tend to angle the knife at approximately 30° to the surface of the curds and make cuts one way across the columns of curds and then back the opposite way, each time going deeper with the knife as I move along the curds (pic 5).

Leave the curds and whey in situ until the pH has dropped to 4.6 or lower. You'll also witness more whey separating from the curds. Gently stir with your hand to check that the pieces are evenly sized; this also enables you to assess the texture of the curds (pic 6).

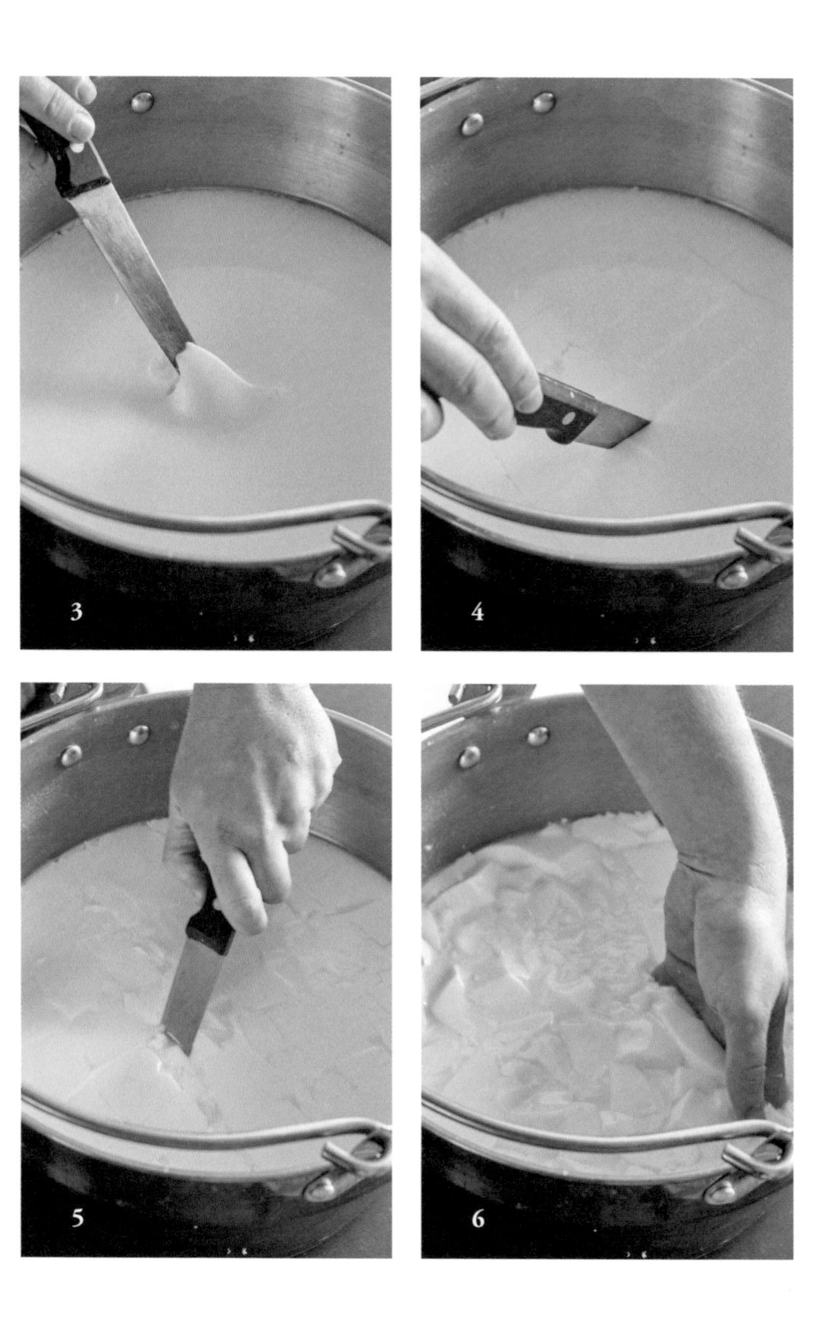

3
4
5
6

Moulding

The next step is to pack the curds into special moulds with drainage holes. Some curds are heated at this stage to create hard cheeses, such as Cheddar, but most soft cheeses are left to drain off their whey naturally. It is important to pack as many curds into the mould as possible, in order to achieve a good-sized cheese (pic 1).

The cheese will need to be left to drain off the whey (pic 2) for between an hour and a day, depending on the type of cheese. To facilitate drainage you may need to turn the cheese: invert on to your hand (pic 3), lift off the mould (pic 4), then place the cheese back in the mould (pic 5) and leave it to drain fully (pic 6).

Typically, curds for hard cheeses are placed in moulds that don't allow free drainage, so the curds are cut smaller to release as much whey as possible before going into the moulds. More whey is expelled as the cheese is pressed into shape in the mould, and the longer maturing time means hard cheeses end up with less moisture than soft cheeses.

The texture of a cheese is largely down to the amount of calcium that remains in it after moulding. If the curds were placed in a mould and drained of the acidic whey very soon after cutting, the acidity of those curds would be quite low. The lower the acidity, the more stretchy calcium is retained within the curds. Soft or semi-soft cheeses, such as Brie, Camembert and raclette, are pliable and stretchy because of this.

3
4
5
6

Ageing cheese

Ageing – also known as maturation or ripening – is part of the journey for many cheeses, allowing them to develop a complex array of flavours and aromas. Unlike fresh cheeses, which should be stored in the fridge at 4°C, and eaten within a few days of making, matured cheeses benefit from a ripening period at a slightly higher temperature. It might be as little as 7 days for a home-made Brie, or 2 years for a Parmesan. The longest maturation time for any cheese in this book is 10 weeks. Understanding the process of ageing is very useful if you make your own cheese – or if you're simply a keen cheese-buyer. Not all cheeses are at their peak when you buy them – a little careful ageing at home can bring them on nicely.

The term for the management of the cheese maturing process is *affinage*, which means 'refinement'. It's not uncommon for professional *affineurs* to take in immature cheeses made by other producers, in order to 'finish' them to perfection. The role of the *affineur* is to create the best environment for the cheese in question. Typically, ageing is all about trying to reduce the amount of moisture in the cheese so that it ends up with the right texture – whether soft and gooey for a Brie, or firm and crumbly for a hard Cheddar.

Certain conditions help a cheese to cross the finishing line. Once the majority of whey has been drained and the cheese has been removed from its mould to start ripening, it should be turned every other day to make sure that each surface of the cheese gets to breathe. This will allow even mould-ripening and avoid sagging in the finished cheese. (Commercial cheeses are 'ironed', meaning a small, cylindrical sample is taken to assess the way they are maturing. It is only cheeses that mature for 6 months or more that require such investigation, so this is unlikely to form part of your cheese-maturing process at home.) The atmosphere needs to offer a balance of high humidity, low temperature, low light levels and good airflow.

Humidity, airflow and temperature

Humidity is measured as a percentage, which reflects the amount of water droplets in the air. Typically, cheese is best matured in humid conditions, somewhere around 90%, with an ideal temperature of 10°C (this can be measured with a dual temperature and humidity device called a thermohygrometer).

On a grand scale, ageing is done in underground caves where these conditions occur naturally. At home, it could be a case of putting your cheese in an upturned plastic food container (as shown overleaf). It is important that there is a little airflow in the cheese 'cave' because this will help regulate the humidity. A humidity level of 90% will facilitate the growth of the bacteria that help to mature the cheese. Too much humidity will make the cheese wet and the bacteria will start to spoil the cheese. If there isn't enough humidity the cheese dries out.

It is possible for a cheese to develop without the benefit of airflow, and some cheeses are deliberately starved of air – such as wax-covered cheeses or cheeses submerged in brine. But if you want to create a cheese rind, airflow is needed. Cellars, though they may be easier to come by than caves, are usually places of stagnant airflow and so not ideal for *affinage*. If you can manufacture a small, fluctuating movement of air, either by causing a draught or circulating air with the aid of a desk-top fan, this will improve the conditions greatly. Don't overdo it, as constantly blowing air will dry the cheese out too quickly.

Temperature is easier to control than humidity and airflow. If the temperature drops below the sweet spot of 10°C, it isn't too much of a problem but it will slow down maturation. However, a temperature higher than 10°C is likely to cause bacteria to develop randomly and this could create a bitter taste in the finished cheese. (Blue cheeses are an exception, as 4°C is the temperature best suited for mould development.)

Your cave A real cave is rarely accessible to the home cheese-making enthusiast, but luckily there are excellent alternatives. Ageing cheese in a plastic box is perfectly fine for the majority of cheeses in this book, but for the more mature, hard cheeses a pantry or cellar should be sought.

There is an area outside my back door that is covered but open ended. This is my 'cave'. I wrap my cheeses either in muslin or breathable, porous paper (two-ply cheese wrapping paper) and put them into an old wooden meat safe (that I rescued from a local car-boot sale) in my 'cave', until they are ready. You can replicate this set-up in a shed, outhouse or garage, as long as the cheese is protected from direct sunlight and safe from flies.

At River Cottage, we have a walk-in chiller that isn't switched on, except for the fan, and this stays at around 10°C. If you have the luxury of a spare fridge then there are ways of hacking the temperature controls by fitting a device that makes them run at higher than normal levels.

I am a great believer in the natural sanitising properties of wood, such as pine, oak and maple, but plastic or stainless-steel shelves for ageing your cheese are perhaps easier to scrub clean. Solid shelving, rather than slatted, is best. Take care to clean the shelves regularly. I clean my wooden shelving after each new cheese has finished maturing, using hot soapy water, and dry it off with a clean tea towel. I don't use anti-bacterial sprays because of the indiscriminate way they kill bacteria – even the beneficial ones.

Salting

I hold salt in high regard. Of course, in excess, it will have a negative effect on your health. However, if used well, good-quality salt has fantastic properties – not least for cheese-making. Cheese cannot be aged without salt. Its ability to remove moisture via a process known as osmosis renders the maturing curds safe from deteriorating bacteria and allows the cheese to ripen. Add salt to natural yoghurt to make labneh (see p.36), a very simple cheese, and you can witness its powerful moisture-drawing effect.

As well as firming the 'paste' of a cheese and enhancing its flavour, salt applied to the outside helps to form a rind, arresting the growth of bad bacteria and actively promoting desirable bacteria and moulds, such as the white bloom of *Penicillium* covering a soft Brie.

The best salts to use For cheese-making, I recommend using the 'gourmet' sea or rock salts. Flaky and crystalline sea salts, such as Maldon, Cornish Sea Salt, West Atlantic Irish Sea Salt or Halen Môn, are evaporated from sea water and have their natural mineral content intact. Rock salts are mined from underground seams of sodium chloride deposited by 'ghost oceans' that no longer exist. Choose a pure salt without additives. Cheap table salt is undesirable because it includes anti-caking agents that stop the salt from attracting moisture – the very thing that it does best! Some table salts are also iodised; iodine has sterilising properties that may render your cheese void of beneficial bacteria.

I collect salts like some people collect ornaments and I have examples from all over the world. No matter if I am using sea or rock salt, for cheese-making I always go to the trouble of grinding the salt with a pestle and mortar, so that it is either easily integrated into the curds or evenly distributed on the outer surface of a cheese; the ground salt can also be used to make a brine.

Brining The brining technique involves either submerging the cheese for a specific time in brine or individually wiping it on a regular basis with a clean cloth doused in brine. You can make a brine by simply adding salt to water or fresh whey. Washing the surface area of a maturing cheese with a brine, or immersing it in the solution, suppresses fungal growth while creating the perfect environment for a complex eco-system of positive bacteria, which both protect the cheese and help develop flavour.

The beneficial yeasts and bacteria that form after brine washing are called *Brevibacterium linens*. They create pungent aromas, which by a twist of taste go hand in hand with extremely rewarding and agreeable flavours. Brine-washed cheeses tend to have rinds in an array of wonderful natural colours that range from pale orange to light pink.

For the wiping method, it is important that to start with the cheese is rubbed at least every other day, to build up a natural defence against fungal growth. The rubbing must be very thorough to ensure every nook and cranny is coated in brine. Depending upon the size of the cheese, it will take around a month of regular washing before the rind pigmentation starts to change and at this point it isn't necessary to wash further; you can allow the cheese to develop on its own trajectory to stinky deliciousness.

The prospect of naturally promoted *Brevibacterium linens* (often truncated to *B. linens*) on the rind of your home-made cheese may give you a sense of unease because of a lack of control over the final product. If freewheeling cheese production goes against the grain, then you can use commercially-made, freeze-dried *B. linens* cultures to help ripen your cheese. These will give you an element of control but, because they contain just a single species type of *B. linens*, they won't give a diverse range of flavours.

Any cheese can be washed in brine, including Cheddar, Gouda and Alpine-style cheeses. However, the overall effect is most striking in the small, soft, washed-rind cheeses that most commonly receive the treatment. (This is due to their higher moisture content and acidity.) As well as producing a colourful, smelly exterior, the bacteria consume the lactic acid in the ripening curds. When the curds have become more alkaline, they release calcium, which eventually causes the paste to 'melt'. Brie is a classic example of this.

How much salt to use The default ratio for salting is 3% salt to the weight of the cheese or fresh curds. This is the minimum amount of salt required to create a preservative environment for the cheese. Anything less will really only be a seasoning: the maturation process might be shortened or the lack of salt might promote the wrong type of exterior 'bloom'.

I also use the 3% rule when I am creating a brine, measuring the weight of salt (in grams) against the volume of whey (in millilitres). I sometimes go to the extreme of making a 'saturated' brine, which means around 28% salt, the point at which the whey can absorb no more salt. I use this to get salt into a cheese really quickly, as opposed to over days, and in particular I prefer this method for creating blue-veined cheeses. (The short, sharp blast of salt punctures the cheese paste, leaving pockets for the *Penicillium* mould to develop, and these are further enhanced by needling; see p.111.)

Brines for washed-rind cheeses Traditionally, a brine for washing will be 15g salt per litre of whey. Because it will already contain beneficial bacteria cultures, whey is a particularly good base, while the salt will activate the right change in acidity on the rind. You can alter the salinity to get different results as the cheese matures.

Doubling the amount of salt to 30g per litre will cause the rind to veer more towards the orange end of the colour spectrum. while weaker brines produce pinkier rinds. The saltier brines tend to reduce the pungent aromas of the finished cheese – this won't necessarily affect the flavour but the saltiness will obviously be more pronounced.

Brines based on beer, cider or wine are very common and require the same amount of salt to liquid as whey. However, some beers, wines or ciders contain a large proportion of residual sulphites, which are left over from the production process. Sulphites used to suppress unwanted yeast growth in wine-making and brewing can slow the growth of beneficial microorganisms on cheese. Organic wines and beers may contain fewer sulphites. Spirits are also a great way to brine-wash cheese – as with classics such as Époisses – but they should be diluted to a ratio of 1 part spirit to 4 parts whey before the salt is added.

Other ripening methods

Techniques such as waxing or cloth binding inhibit surface ripening, slowing down the ageing process and allowing the cheese paste to develop over a long time period. Some Cheddars are matured this way. Ash, leaf and birch bark are also excellent for wrapping around soft cheese and will bring extra flavour, as well as keeping it moist and protecting the cheese when it is handled.

Mould-ripening is another method applied to both soft and hard cheeses; this is covered in more detail in Matured Cheeses (see pp.108 and 124–33).

Cold smoking is a technique that not only gives additional flavour to a cheese but also helps to dry and preserve it. This means it's a technique suitable for hard cheeses only.

Cheese-making Ingredients

The best home-made cheese products are created using the highest quality ingredients; if you think about it, there aren't many ingredients required to make cheese, so they may as well be the best. You can follow the tried and tested path when it comes to adding commercial starter cultures (which affect the flavour) and rennet (which mainly affects the texture) or you can be a 'guerilla' cheese revolutionist and create more maverick, naturally inoculated cheeses using a kefir starter culture and vegetable rennet. Either way is valid but I would suggest learning your craft using the conventional method before moving on to the latter.

Choosing milk

The cheese recipes in the chapters that follow can all be made with either raw or pasteurised milk. Raw milk has more complex flavours, coagulates more easily, is more closely aligned with tradition and regionality, and is now favoured by a new wave of independent cheese-makers (see p.24). Pasteurised milk, on the other hand, still offers a broad range of cheese and dairy opportunities, with the added benefit of greater security and control over the process and the outcome (see p.23).

Using raw milk for home-made cheese is the extreme end of the craft. Raw milk can harbour pathogenic bacteria such as listeria and spoilage organisms called 'pseudomonas' even when it is stored correctly in a fridge. Having said that, there is a difference between the risk attached to raw milk, and the risk attached to products made from it. The bad bacteria found in raw milk do not like dry, acidic and salty conditions; this is exactly the kind of environment certain cheeses provide. The harder and more mature the cheese, the less hospitable it is to pathogens – whether it was made with unpasteurised milk or not. Soft, young cheeses carry a higher risk of pathogens because of their moist and more alkaline nature; but pasteurising is not necessarily the benchmark for safety here because milk or cheese can be contaminated after pasteurisation.

For cheese-making on a domestic scale, the use of raw milk is challenging because it is not possible to monitor or control bacteria while the cheese ages. However, I do use raw, local milk for some of my dairy products, especially Caerphilly. If you want to do the same, ensure your milk comes from a reputable, traceable source, be strict about cleanliness and hygiene, and follow recipes carefully. The increase in acidity in the early stages of cheese-making is the crucial point: if the acidified milk reaches a pH of 4.6 or below, you can be confident that pathogens won't be able to survive.

If you want to buy raw milk, there are licensed farms that sell directly at the farm gate, at farmers' markets or via co-operatives that will deliver to you. These farms must be open to stringent analysis of their process and products by local health authorities and the Food Standards Agency. Raw milk should only come from a closed herd that is TB- and disease-free. I take it as a positive sign if the farm also makes their own raw-milk produce (cheese, butter etc.), because it shows a level of confidence in the product and indicates that a rigorous testing procedure must already be in place. This is particularly the case if the farm produces soft cheese from raw milk – this is made at relatively low temperatures with very little time to ripen or dry out. Their controls must be spot-on to ensure safe practice.

Once you've tackled the issue of 'raw v. pasteurised' you can look at other factors when choosing milk. Certain products require a specific animal milk – such as *chèvre* cheese being made from goat's milk, or mozzarella from buffalo milk. Some cheeses can be made from any milk – cow's, goat's or sheep's. However, the time required for the curds to set will vary. Sheep's milk sets much more quickly than cow's milk, whereas goat's milk is very slow to set. Most of my recipes assume that you are using cow's milk.

A higher percentage of fat means better curds. Most cheese-making recipes suggest using whole milk, rather than skimmed or semi-skimmed. (However, it is possible to make cheese from lower fat milks, such as my own county's glorious Dorset Blue Vinny, for which the milk is hand skimmed to deliberately reduce the fat content.)

The fat content of milk is not altered by the pasteurisation process but it is important to know that homogenised milk – which represents the vast majority of milk sold in the UK – is not ideal for cheese-making because it does not have the capacity to produce strong curds. Fortunately, you can buy pasteurised, unhomogenised milk in some supermarkets and farm shops, and it can even be available via doorstep delivery. (See p.28 for more on homogenised milk.)

Freshness is important. If you have access to a farm that milks, pasteurises and bottles on site, and can buy that milk on the day of production, then your chances of making a good cheese are improved; even the best organic whole milk from a supermarket will already be 3 days old.

UHT milk has been raised to a temperature above boiling point. This kills all the bacteria and while that means the milk can be kept out of the fridge almost indefinitely, it also spells the end for it as a cheese-making ingredient. The chemistry of UHT milk is so radically different from raw milk that it just will not form curds and cannot be used to make cheese.

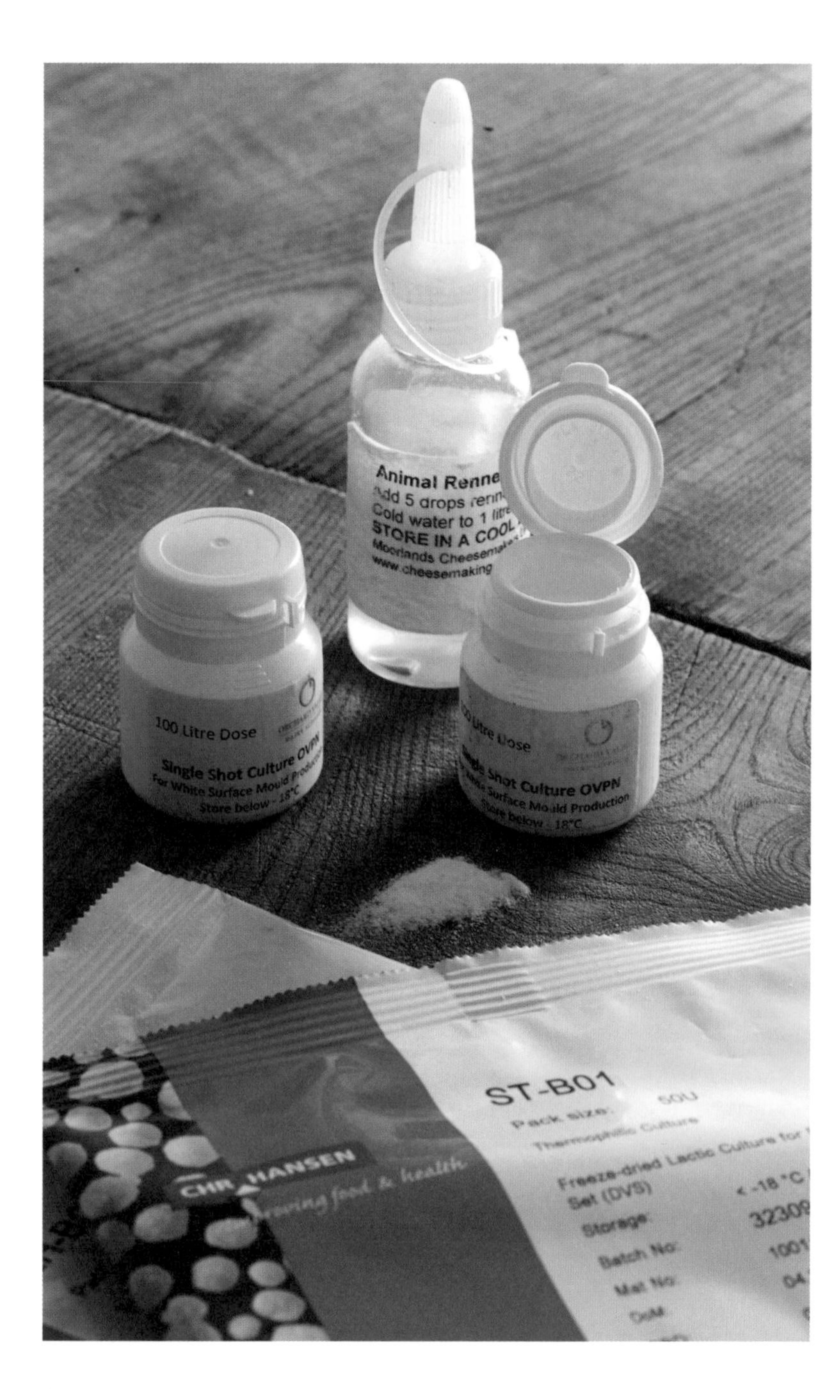
Animal Renne
STORE IN A COOL
100 Litre Dose
Single Shot Culture OVPN
For White Surface Mould Production
Store below - 18°C
ST-B01
Pack size: 50U
Thermophilic Culture
CHR HANSEN
Set (DVS)
Storage:
Batch No:
Mat No:

Starter cultures

All cheeses begin with the acidification of milk, which causes it to split into solid curds and liquid whey. If milk is left to its own devices, it turns (acidifies) through the activity of the *Lactobaccillus* bacteria naturally present in it. This is often called the souring, clotting or 'clabbering' of the milk; it is actually milk turning into cheese in an uncontrolled manner.

The careful control of acidity in the first stage of cheese-making is important for driving off unwanted, harmful organisms. Most cheeses are acidified by adding starter cultures, which are made up of carefully selected good bacteria, to warmed fresh milk. I tend to pour on the starter culture and let it settle on the surface of the milk for a moment before stirring it in. This allows the culture to rehydrate and become active, causing the acidity level of the milk to rise as it ferments.

For the novice cheese-maker, commercial starter cultures are the best place to begin and the majority of the recipes in this book use them. These cultures are available online from specialist suppliers (see the Directory, p.218). I have also included a method for making your own starter culture. This is not for the total beginner but once you have got to grips with the cheese-making process, it gives you the option of making a truly territorial cheese, infused with microbes that you have grown in your own domestic set-up.

With my first attempts at making cheeses, I always used commercial cultures and pasteurised milk. As I gained experience, I developed the confidence to move towards a more relaxed, natural approach that sometimes also incorporated raw milk and kefir-inoculated whey starters, which ultimately led to some of my favourite outcomes.

Commercial starter cultures are usually made up of specific strains of lactic acid bacteria (LAB) with a distinct set of characteristics that determine how a cheese will develop. These cultures sometimes also contain yeasts and moulds that have a role to play along the way. Starter cultures are made by just a few companies and the two most popular brands are Christian Hansen and Danisco, which is a subsidiary company of DuPont. Far and away the most popular starter culture among home-made-cheese enthusiasts are the ones categorised as Direct Vat Inoculation (DVI) cultures.

The most commonly used DVIs are freeze-dried and the 'living' cultures that they contain start to become active once they are stirred into milk at the requisite temperature. They usually come in small packets or sachets and tend to be better value for money if you buy them in batches. With DVI starters, the calculations of how much to add and at what temperature, in order to best activate the properties of the culture, are usually laid out for you. DVI starters can be stored in the fridge or freezer for several months.

Other cultures, which are natural but still processed, are usually confined to use within the artisan cheese-making community, although they are available commercially. They include 'mother culture,' which is derived from raw milk that has naturally acidified, 'bulk starter', which is a mother starter that has been frozen into individual pellets, and 'whey starter', which is made from the whey of the previous day's cheese-making.

Calculating the weight of your starter

In the cheese-making recipes in the following chapters, I will tell you how much of a particular starter to use (and sometimes it is worth buying a particular brand too). I will give the weight in units, which then need to be converted by you into grams. If instructions are not included with the starter culture when you buy it, you can use a calculation that comes from cheese-making expert Paul Thomas. Paul's equation is as follows:

$$\text{Weight of starter you need} = \frac{\text{Units of starter given in recipe} \times \text{Total weight of starter in sachet}}{\text{Total units in sachet}}$$

So, if a recipe calls for 2 units of starter culture and the sachet contains 50 units and weighs 10.2 grams the calculation would be:

Weight of starter required = 2 x 10.2g ÷ 50 = 0.408g

Using digital scales which are accurate to 0.01g, this would be rounded up to 0.41 grams.

Mesophiles and thermophiles

Some starter cultures are predisposed to thrive in warm conditions: these are called mesophiles and their preferred temperature range is 20–40°C. Others, namely thermophiles, are more effective in a hot environment; their preferred temperature range is 30–55°C. At the start of the cheese-making process, milk is heated to different temperatures according to the type of cheese, which determines the choice of starter culture. In theory, you could use a mesophile starter culture at a high heat and it would still have some positive effect, but a thermophile starter culture would be more effective.

Within the group of mesophilic starter cultures, there are two subdivisions: heterofermentative and homofermentative. These will have different effects on the cheese. Both produce lactic acid, but homofermentative starter cultures also create gas and aroma compounds. Homofermentative starter cultures are generally used for making harder cheeses.

Yeasts and moulds

The initial requirement for any starter culture is to quickly raise the acidity of the milk (i.e. *lower* the pH). This acidic state stops other, potentially hazardous bacteria growing. The yeasts and moulds sometimes contained within the starter then get to work as the cheese ripens and matures. The type of yeasts often present within a DVI starter culture, such as *Geotrichum* and *Debaryomyces*, are chosen for their ability to quickly consume lactic acid. This increases the acidity, which is good because the enzymes that stabilise the cheese throughout the ageing process thrive in acidic, humid environments. The yeasts can have an effect on the texture of the cheese – particularly on the surface, where they leave wrinkles and rolls (as in the wonderful Somerset Stawley cheese, which is made from goat's milk).

Moulds contained within a starter, such as *Penicillium candidum* or the rather unappetising sounding *Mucor*, will eventually make their appearance on the surface of ripened cheeses and can appear within the paste of the cheese as well. Blue *Penicillium roqueforti* is the most common mould you'll see in the paste – the cheese needs to be 'needled' for the mould to appear, creating air gaps where it can take hold. Moulds have an effect on the taste of the cheese, as well as giving a distinct appearance. They are responsible for the white powdery covering on soft cheese such as Camembert or Brie, as well as the blue veins in a Dorset Blue Vinny.

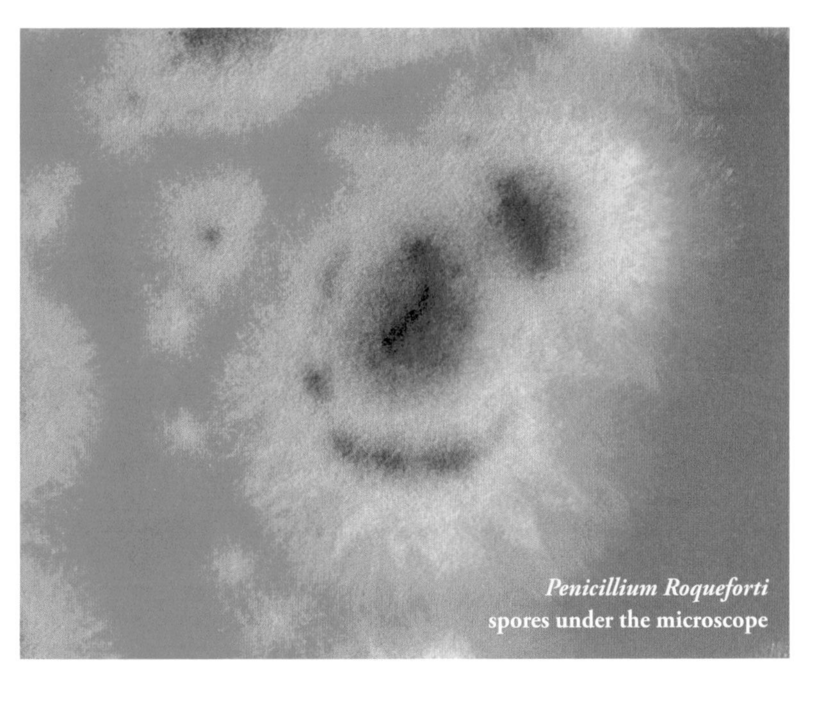

Penicillium Roqueforti
spores under the microscope

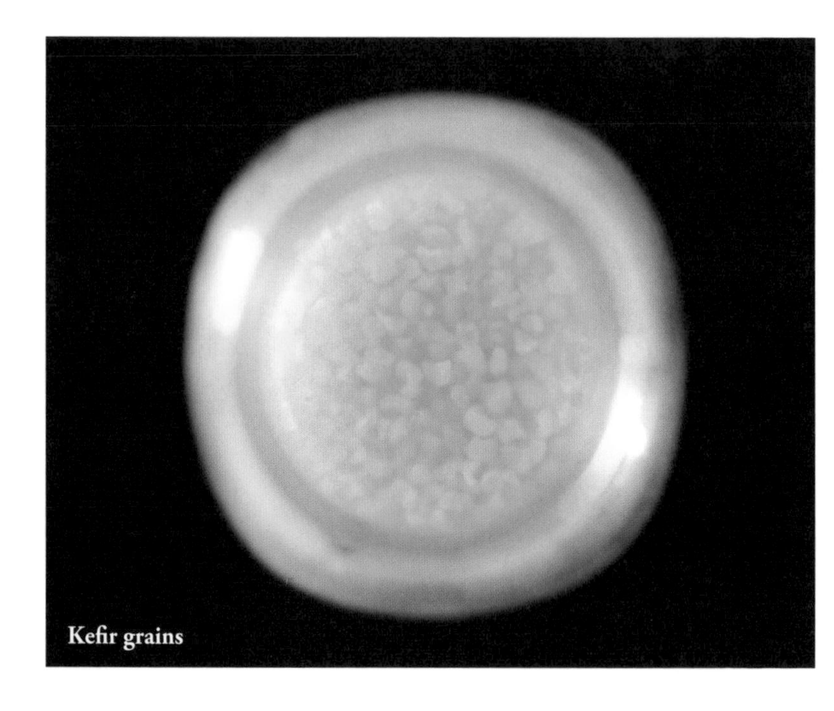

Kefir grains

Making kefir starter

Kefir 'grains' are an ancient mystery. They originated in Central Asia thousands of years ago, and have enjoyed a renaissance recently. They are not grains in the sense of being plant seeds, but rather colonies of yeasts and bacteria that form small, white granules. These granules clump together to look like little cauliflower florets, or cottage cheese, and if you add them to milk, they cause fermentation and thickening. The resulting liquid, which is itself called kefir, contains numerous beneficial bacterial cultures and fungi which have a probiotic effect on our digestive systems. Many of the bacterial cultures in kefir are similar to those found in your digestive tract and can benefit your skin, teeth and general wellbeing – kefir's literal translation is 'good feeling'. Kefir grains are widely available in health-food shops and online.

The bacteria found in kefir closely resemble those found in raw milk and so kefir can be used in cheese-making. Using a natural starter such as kefir is exciting because it allows you to step outside the fairly narrow realm of commercial starters. It does, however, require a certain confidence. You have to be comfortable with the notion of moulds and bacteria that have not been synthesised in laboratories and are part of the natural and therefore unpredictable microflora around us. Using kefir to make cheese feels a bit like the first time you ride a bicycle without holding the handlebars: if you keep pedalling, the bike still goes, but you don't have too much say in the direction. It's the same with kefir; the acidification and curdling will happen but the cheese will be something of an unknown quantity. Although you may not necessarily have the dead cert result you would get with a commercial culture and there is inevitably an increased risk of pathogens developing (see the Troubleshooting guide on pp.208–13), a kefir culture can give depth of flavour to pasteurised milk cheese that would normally only be possible with raw milk.

I suggest that you only opt for a natural kefir starter after you have got used to working with commercial starters. You will gain experience and assurance from the more conventional method that will stand you in good stead should you decide to become a guerrilla cheese-maker.

In order to harness the goodness found in kefir grains you must take on the small responsibility of keeping them alive with regular feeding. It is quite difficult to kill them off completely but if they are to be the bacterial engine of your non-conformist cheese then they should be treated with respect. All that kefir grains need in order to work their magic is milk. It doesn't even matter which type, though you'll find that different kinds of milk – raw, or pasteurised whole (full-fat) milk, taken from cows, sheep, buffalo or goats – will produce different textures and tastes. Feeding kefir is much like keeping a sourdough bread starter alive.

Just 1 tsp (5ml) kefir grains added to 240ml milk and left at room temperature for 24 hours will be sufficient to create a starter culture. The milk should have thickened and become slightly (I would say pleasantly) sour and acidic.

You will need to strain the kefir before using it – don't discard the grains, just set them aside for the next batch of kefir; as long as they are kept dry in an airtight container in the fridge, they will remain alive but dormant for months.

For every 4.5 litres milk you're using to make your cheese, you will require 60ml kefir. This is added at the stage when you would usually use the commercial starter culture. (There will be lots of kefir left over; you can either keep this to make more cheese, or simply drink it.)

P.S. If you are making a mould-ripened cheese, leave the kefir grains in the milk for longer, at least 48 hours, to allow the yeasts and mould naturally present in the kefir to develop. This more potent mix can then be used as above.

Rennet

You can make cheese from milk that is purely acid-curdled (see Paneer, p.91) but it is more common and efficient to use a starter culture and rennet. Effectively a catalyst, rennet is an enzyme that speeds up the coagulation process begun by the starter culture. It makes the curds stronger, more rubbery and easier to handle.

Rennet can take several forms. Traditionally, it is an enzyme called chymosin, found in the fourth digestive compartment (abomasum) of a ruminant's stomach. In cheese-making, the chymosin of young cows, goats and sheep is most often used. Originally, segments of the abomasum would be placed in the milk to make it curdle and this method is still in use today, especially in Pecorino-style cheeses. But generally the chymosin is extracted and formed into a liquid, powder or tablet.

Some forms of rennet can also be harvested from vegetation and fungi, and are suitable for vegetarians. Rennet from natural fungal cultures is often referred to as a 'microbial rennet'. Cardoon stamen, artichoke thistles and figs also produce enzymes that can be used in the same way as animal rennet. The only issue with plant-based rennets is that they can cause bitter notes if too much is added initially and there is residual rennet left in the ripened cheese.

As a cheese matures, the rennet it contains helps to break down proteins in a process known as 'proteolysis', which develops flavour and aroma. Another enzyme responsible for developing flavour is lipase. This is also present in the abomasum, and in small quantities in milk itself, especially in raw milk. Lipase varies in strength from animal to animal and is mildest from a calf, intermediate in kid goat and strongest in lamb. Lipase helps break down the fat in milk and is particularly valued for making long-aged cheeses.

The ethics of rennet

As I have already said, there are plenty of issues in the dairy industry regarding animal welfare. There is also some stickiness surrounding rennet. It is a by-product of the wasteful slaughter of very young male dairy calves. Although there is an increase in rearing these calves for meat (usually rose veal), more often than not they are culled shortly after birth. The extraction of rennet is one of the few uses to which these carcasses can be put before they are destroyed.

The growing popularity of vegetable or microbial rennets helps bypass this issue – but creates another one. Some rennets made from a non-animal microbial source can contain genetically modified cultures – check the ingredients carefully if you are worried about this. Happily, it is possible to make exceptional cheese using organic vegetable rennet. Some cheese-makers regard vegetable rennet as lacking in strength compared to animal rennet, but I'm more than happy to use it. You can also produce vegetable rennet at home (see p.76).

Using rennet

The rennet is added to the milk about an hour after the starter culture. Having briefly stirred in the rennet, you can step back and allow coagulation to take place; your only task now is to maintain the minimum temperature of the milk. Flocculation (the first signs of coagulation) can be reached in as little as 15 minutes, but speed isn't necessarily a benchmark of success. Occasionally, the freshness of the milk or the fat content has a bearing on the coagulation time. Just relax and let it happen. If you add more than the specified amount of rennet in a bid to aid coagulation, you run the risk of altering both the flavour and integrity of the finished cheese.

Buying rennet

You can get different strengths of rennet; the really high strength versions are only used on an industrial scale. Rennet strength is measured in International Milk Clotting Units (IMCU). The most common IMCU strength in the UK is 1:10,000 and this is the strength used in all of the recipes in this book.

Rennet is available as a liquid, powder, tablet or paste. I tend to use liquid rennet but the various forms are interchangeable, and you can use whichever you like for my recipes, as long as it is efficient at the temperature required for the cheese.

All forms of rennet are easy to use. Some people suggest that you should dissolve the rennet in a small amount of water before adding it to the milk, others that you can add it directly. I have tried both methods and haven't found either one to be more successful than the other. However, if you prefer to dissolve the rennet first, then liquid rennet is perhaps the most practical choice as it is easy to measure and dissolves instantly.

Powdered rennet (which is made from purified, dried chymosin) often contains additional preservatives, including salt, which is good for shelf-life. The downside is that the sachets of powdered rennet tend to be large and if you are making cheese on a domestic scale, you'll need to calculate and weigh a smaller quantity with the help of some seriously accurate digital scales.

Tablet rennet is more geared towards the small-batch cheese-maker. The tablets are made of compressed powder rennet, in manageable small quantities.

Paste rennet is simply the salted and dried abomasum, ground into a paste. It is quite rare and is generally used only by a small circle of guerrilla cheese-makers looking to pare back the process of cheese-making to its barest essentials. The benefit of using this least-adulterated form of rennet is that it contains large quantities of lipase, which helps develop the taste of a maturing cheese.

How to make a vegetable rennet

There are two wonderful cardoon plants in the River Cottage HQ kitchen garden. Majestic and statuesque, with a slightly prehistoric air, they dominate the herb borders adjacent to the farmhouse and up until recently I have been an admirer of their giant thistly presence without ever thinking of them as ingredients. However, I then discovered that their purplish flowers are excellent for coagulating milk and they are used to make some traditional cheeses along the Iberian Peninsula, such as Serra da Estrela.

The coagulating properties come from an enzyme within the stamens of the flower. These need to be dried and pulverised before use. Then a decoction is made: 2g pulverised dried cardoon stamens are steeped in 60ml warm water for 1 hour. The strained-off liquid can now be used as a replacement for animal rennet. There will be enough coagulant from these quantities to inoculate 4.5 litres warm milk. If the process seems a little too involved, I recommend buying a good vegetable rennet instead.

Dried cardoon plant

Equipment

Most of the kit you need to make your own dairy products is likely to be in your kitchen armoury already – or can be improvised from it. However, purpose-made items are readily available and give you more control over, and confidence in, the whole process. For some dairy products, certain equipment is important. For example, while a home-made butter requires only a bowl and a whisk, a hard Cheddar-style cheese needs a good set of digital scales, a thermometer, a pH meter and cheese moulds. Here is a guide, with low-fi alternatives suggested where possible.

Cheesecloth and muslin

These inexpensive, loose-woven cloths are used for draining fresh cheese curds or yoghurt. The only difference between them is that muslin is a little finer than cheesecloth. Some cheeses are drained in a mould lined with cheesecloth or muslin to reduce the amount of whey lost. You can buy disposable, single-use cheesecloths, which are usually blue in colour, or more traditional cotton cheesecloths or muslin, which require cleaning between uses.

A muslin draining bag works extremely well for quick curd cheeses and labneh: you put the fresh curds in, then draw the strings of the bag together tightly, which accelerates the expulsion of whey through the fabric.

Draining mat

While your curds are draining in their moulds, you will need to stop the released whey from going everywhere. It is worth saving the whey, which can be used for brine (see p.62), whey butter (see p.105) or ricotta (see p.102). You can sit your cheese moulds on a metal rack over a tray, but commercial draining mats made from plastic mesh set over a plastic container are a better size. They usually have enough space to accommodate around 10 moulds.

Hands

Obviously, these are essential tools. Sometimes there is no substitute for the sense of touch. Immersing your hands in a bowl of recently cut, warm curds enables you to judge firmness as well as temperature and texture – it's a pleasing experience. I like to transfer the curds into cheese moulds by hand, and press them down with the backs of my fingers. And you rely on the dexterity of your hands to flip a cheese in its mould.

The act of salting a cheese rind or rubbing it with a brine wash keeps you in contact with the cheese as it develops from fresh to mature. The sharp tap of a finger on the outer rind of a hard cheese can tell you a lot about how that cheese is ageing too. It goes without saying that your hands, like all other tools used with dairy products, must be very clean.

Moulds

These are perforated vessels for shaping curds into cheeses and they are available in a range of shapes and sizes. The perforations allow whey to drain off.

Typically, moulds intended for soft cheeses tend to be deeper than they are wide, with many drainage holes. They allow the loss of up to two-thirds of the original volume via draining.

Moulds for hard cheeses tend to be wider than they are tall, with fewer drainage holes, which are mostly concentrated along the bottom. They usually come with a 'follower' which fits inside the mould on top of the curds and is used to press out the whey and push the curds into a block shape.

To keep things interesting, I like to keep an array of moulds in different shapes and sizes. Most of mine are made from plastic and, whatever your thoughts are on that, they are cheap and easy to clean. It is also possible to use clay moulds, or baskets woven from reeds and willow. You can improvise your own moulds out of reclaimed plastic pots or baskets; this is something we do at River Cottage HQ from time to time.

pH meter

Acidity is absolutely key when making cheese. Once the milk has been inoculated with starter bacteria and rennet, the acidity will increase, which has the effect, among other things, of seeing off unwanted pathogens. Acidity can also affect the flavour, texture and development of mould within the cheese.

If you use a commercial starter, measure it accurately and follow the instructions carefully, a rise in acidity should be assured. However, it is helpful and confidence-inspiring to be able to watch the acidity rising (shown by the pH falling from the milk's original pH of around 6.5). Litmus paper can indicate this but won't give you an exact reading. I would recommend getting a pH meter, ideally one that can give you readings to two decimal points.

Keep a cheese notebook and record your acidity readings for each batch of cheese. It's useful when comparing finished cheeses to know what the acidity levels were like during their making. If I make a cheese I particularly like, I try to match the pH at flocculation (the coagulating stage) next time.

Preserving pan

Although any large stainless-steel pan or pot will do, a proper, high-sided preserving pan is invaluable for warming a large volume of milk and allowing curds to form. As the ratio of milk to cheese yield tends to be around 10 to 1, you need a pan with a large capacity. Most standard preserving pans have a capacity of 9 litres, which allows you to produce up to 900g cheese in one batch. Preserving pans have strong handles and are well balanced, which is particularly useful for draining whey or

pouring curds carefully into moulds. They are wide, which gives you easy access when cutting the curds with a knife. They also have a thick, heavy base, which ensures heat is retained, stopping the temperature from falling too quickly; this helps to prevent hot spots too.

Scales

Digital scales that measure to an accuracy of 0.01g are perfect for weighing small amounts of rennet or starter (you will seldom use a whole sachet at one time). Choose scales with clear reading windows that can be tared, or zero-ed. This way you can set a container on the scales, then zero them and add the correct quantity directly into the container.

Syringe

If you use liquid rennet, then a syringe is a useful item of kit because, again, you will need to measure very small quantities. A simple medical or food-standard syringe – from a chemist or a cheese kit supplier – is ideal. Choose one with a scale up to 5ml along the side. A syringe is also useful for drawing off whey to measure its pH.

Thermometer

Dairy products have been made for many millennia without thermometers but they should nevertheless be part of the modern enthusiast's kit. There are moments in most dairy recipes when specific temperatures matter and a thermometer offers a level of precision and therefore security in these crucial early stages.

Get the most accurate, high-quality thermometer you can. My favourite is the hand-held digital probe thermometer (shown opposite). I use this while heating milk, to take regular readings in several places around the pan, rather than leaving the probe suspended in one place.

Utensils

Keep an array of colanders, sturdy sieves, slotted and perforated spoons, ladles, whisks and measuring jugs within easy reach. For all but the simplest of cheeses, you'll also need a long, thin, sharp knife for cutting the curds. There are specialised curd knives on the market but I find a flexible, thin-bladed kitchen knife works just as well.

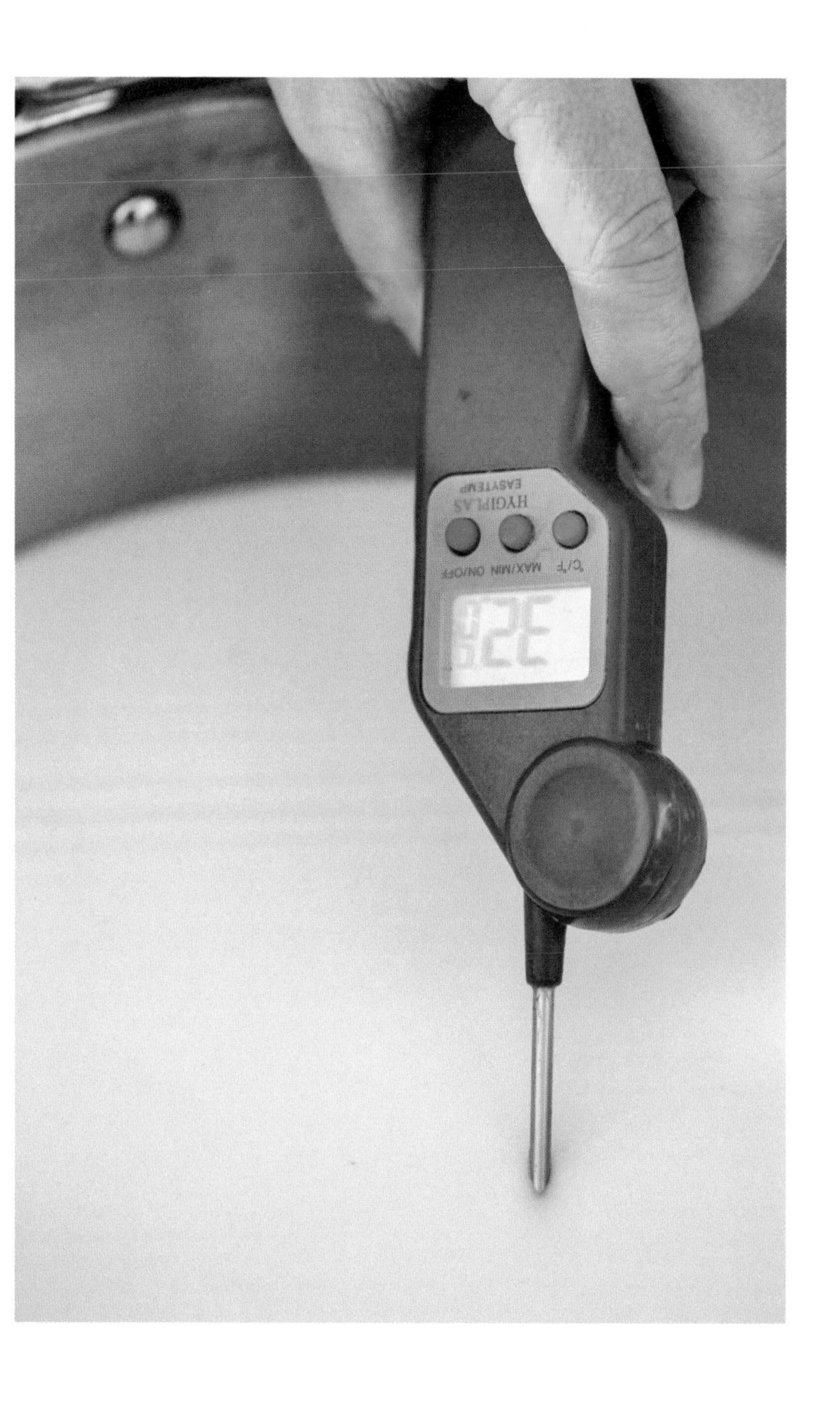
HYGIPLAS
EASYTEMP
ON/OFF
MAX/MIN
°C/°F

Fresh Cheeses

Fresh cheeses are edible as soon as they are made; i.e. they do not require a subsequent maturing or ageing process. This means they have a very limited shelf-life (unless, of course, preservatives are added, as they are for many commercially produced fresh cheeses). Most fresh cheeses are fluffy and soft with a simple taste, which may be salty or tangy, often with citrus notes. They can be flavoured with extra herbs, garlic, peppercorns or even ash. Fresh cheeses are wonderful in their own right and they're very quick to make; they are also a great stepping stone on the way to making more involved, renneted cheeses.

The basic process is the same for making all fresh cheeses: milk is soured by adding an acid such as lemon juice, or a starter culture that contains specially selected bacteria. It is this acidification that triggers the curds to form. Fresh cheeses are generally made at lower temperatures than renneted cheeses and they carry a high amount of moisture, so you need to be scrupulous about cleanliness and achieving the correct temperature at each stage to prevent pathogens developing.

Sometimes cream is added to the acidified curds to increase the velvety texture and fat content, as in traditional cottage cheese (see p.92). Ricotta (see p.102) is classified as a fresh cheese but it is made using heated whey rather than the curds. Mozzarella (see p.96) is also a fresh cheese, made by quickly heating and then cooling and stretching the fresh curds.

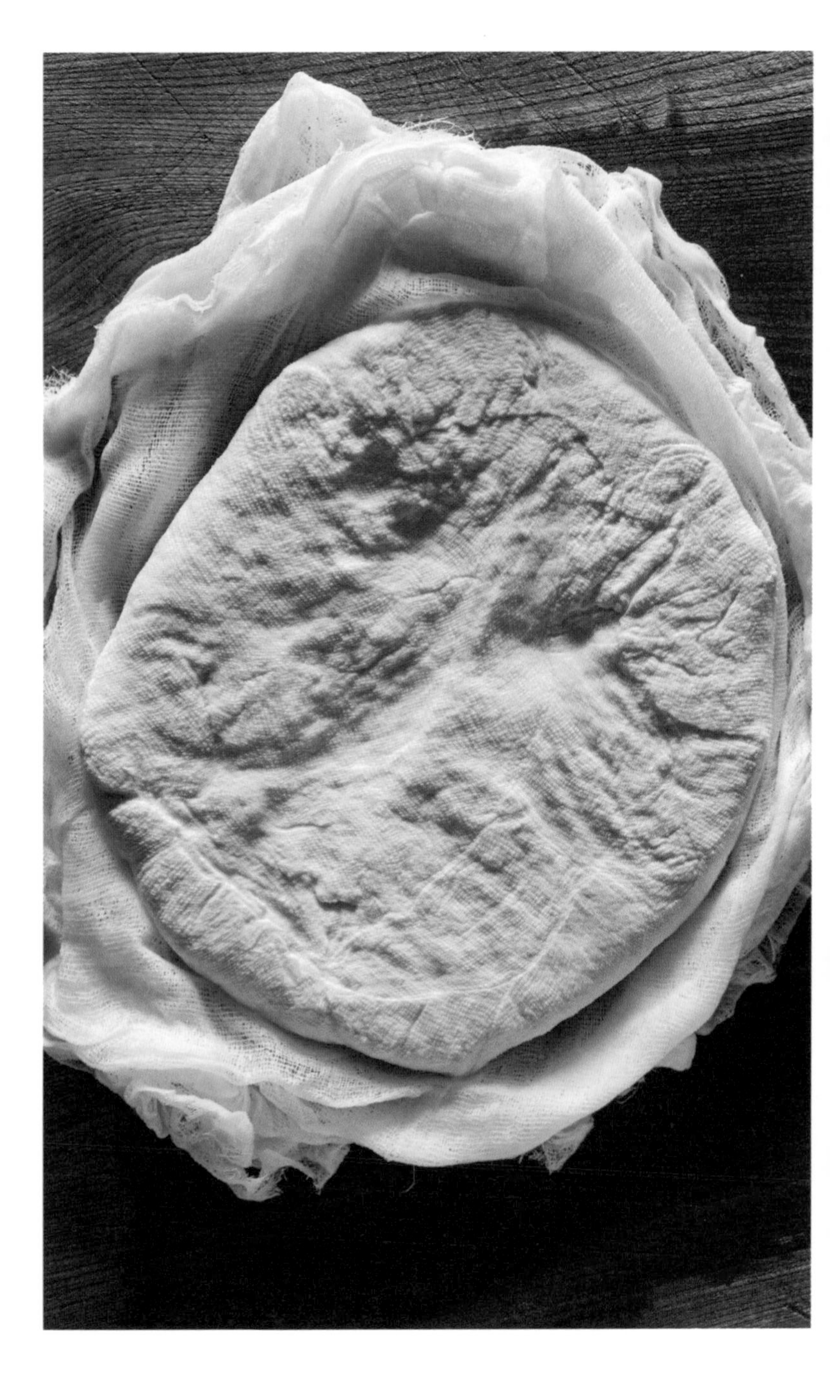

Paneer

Associated with Indian cuisine, this is a 'directly acidified' cheese – lemon juice or vinegar is added to the milk, rather than an acid-producing bacterial culture. Because of this, paneer does not develop huge flavour, but its mild taste and yielding texture are lovely with spicy accompaniments. It is typically used in cooked dishes, where it softens rather than melts. Add it to curries or fry it gently and serve with veg dishes, eggs or tomatoes.

Makes about 500g

5 litres whole cow's or buffalo milk, or a combination (unhomogenised)

150ml distilled white vinegar or lemon juice

Pour the milk into a preserving pan and heat it slowly until it reaches 90°C, stirring every so often with a wooden spoon. Take regular readings in several places around the pan, rather than leaving the thermometer suspended in the milk in one place. Remove the pan from the heat and stir in the vinegar or lemon juice.

Allow the milk to cool. Within 10 minutes, it will begin to coagulate into small curds. As the curds form, they will sink to the bottom of the pan.

Line a large sieve with muslin or cheesecloth and suspend it over a bowl or the sink. When the whey is free from curds at the top of the pan, scoop out the curds using the slotted spoon and place them in the cloth-lined sieve. It is better to handle the delicate curds gently in this way, rather than just pouring the pan's contents directly into the muslin. Discard the whey left in the pan as, unfortunately, it cannot be used for whey butter or ricotta, as it has not been cultured.

Place a chopping board on a tray or the draining board. Fold in the corners of the muslin over the curds and lift them on to the board. Place another chopping board on top and press down so that the paneer becomes flattened and the excess whey is squeezed out. Place a weight (or a roasting tray filled with water will do) on the top board to keep the paneer pressed. Make sure the boards are steady and won't allow the tray of water, if using, to slip off. Leave for 15 minutes.

Remove the flattened paneer from the muslin. Slip it straight into a bowl of cold water and leave it immersed for about 30 minutes.

Remove the paneer from the water and pat it dry with a clean tea towel. Store, wrapped, in the fridge for up to a week. I refrain from adding salt to paneer until I cook with it, to avoid altering the texture.

Cottage cheese

A real cottage cheese can be deliciously creamy, tangy and rich – a world away from most commercial versions. For the novice cheese-maker, this is an ideal next recipe to attempt after making yoghurt or paneer because it uses a starter and rennet and you can try your hand at cutting the curds. Whole, pasteurised milk will give you great results but you can also use raw milk (see p.24). The cheese is particularly stable (you might say 'safe'), as it is heated to a high temperature quite quickly, acidified with the use of a reliable, commercial starter, and helped to coagulate by the addition of a little rennet. The finished cheese can be enriched with cream if you like. I like to serve it scattered with fresh herbs.

Makes about 500g

9 litres whole cow's, sheep's or goat's milk (unhomogenised)
2 units of homofermentative mesophilic starter (see p.70)
½ tsp liquid rennet (or 10 drops from a pipette)
About ½ tsp sea salt, to taste
300ml double cream (optional)

Pour the milk into a preserving pan and gently warm it over a low to medium heat until it reaches 30°C. Take regular readings in several places around the pan, rather than leaving the thermometer suspended in the milk in one place. Stir the milk with a wooden spoon to avoid it catching and to help distribute the heat. When it reaches the correct temperature, remove from the heat.

Add the starter and stir it into the milk so that it is fully combined and none is left on the surface. Allow the milk to stand for an hour.

Add the rennet and stir in thoroughly. Leave until the curds have set to a firm blancmange consistency – this may take up to 6 hours. You can judge the curds by eye, or check the acidity level using a pH meter – the pH should have lowered to about 4.6.

Cut the curds using the grid pattern technique (see pp.52–3). Place the pan back over a very low heat, slowly stirring the curds and whey until the temperature reaches 39°C. This gentle heating should take place over 30 minutes to an hour. When the target temperature is reached, stop stirring and remove from the heat.

Allow the curds to settle to the bottom of the whey then drain in a sieve; you can keep the whey for another recipe or to make a brine. Run cold water over the curds continuously until cooled to 5°C, then transfer to a bowl and stir in the salt and cream, if using. Serve at once or store, covered, in the fridge for 7–10 days.

Fresh curd cheese

Pure white with a soft, fluffy texture, fresh curd cheese (aka 'lactic cheese') is delicious spread on crusty bread or broken into salads. It can be ready to eat in just a couple of days and is moist and mild, ready to take on extra flavours. Traditionally, ash was used as a way of protecting and preserving this very young cheese. Here, modern cheese-maker's ash is sprinkled over the finished cheese, which lends a lovely mineral flavour and enhances the appearance.

Makes 8 small cheeses (each about 100g)

8 litres whole cow's, sheep's or goat's milk (unhomogenised)
0.6 units of homofermentative mesophilic starter CHR-Hansen R-704 (see p.70)
1ml liquid rennet (measured in a syringe for accuracy)
15g sea salt
Cheese-maker's ash, to finish (optional)

Heat the milk in a preserving pan gently until it reaches 21–22°C. Take regular readings in several places around the pan, rather than leaving the thermometer suspended in the milk in one place. Remove from the heat.

Add the starter culture and stir thoroughly, then immediately stir in the rennet so that it is dispersed evenly. Cover the pan and leave the milk to coagulate slowly at room temperature over a period of 10 hours or so, undisturbed.

When the curds have set into a gel or register 4.6 or thereabouts on a pH meter, they are ready to drain. Transfer the curds by hand to a draining bag (or muslin-lined sieve) suspended over a bowl; they should not be cut at all but just manually broken up and placed in the bag. The slippery, soft, silky curds will be almost solid but with a liquid movement, offering up just a small amount of resistance. There will be some considerable weight in the bag so make sure it is secure.

Leave to drain at room temperature for at least 12 hours, up to 24 hours. During this time, you can massage the bag gently to encourage further draining of whey.

Once the curds have drained significantly so there is no residual whey, open the bag and mix in the salt – or you can hold off salting until the cheeses are shaped.

Remove the cheese from the bag and form it into 8 individual shapes by hand. You can finish them off with a dusting of ash – or a mixture of ash and salt – if you like, or leave them plain. These cheeses can be eaten straight away or stored in the fridge for up to a week.

Mozzarella

Sweet, milky mozzarella can be made from cow's or buffalo's milk. If you can get hold of the latter, your mozzarella will be closer to the real Italian thing, with a superior, silky, elastic texture. Either way, mozzarella is versatile – it can be paired with basil and ripe tomatoes in an Italian caprese salad, tossed into a panzanella salad, scattered on a pizza, torn into a pasta dish or eaten with prosciutto.

Makes 3–5 balls, depending on size

4 litres whole cow's or buffalo milk (unhomogenised), at room temperature
1 unit of *Streptococcus thermophilus* starter (see p.70)
1.5ml liquid rennet (measured in a syringe for accuracy)
Sea salt

Pour the milk into a preserving pan and add the starter while the milk is still at room temperature. Stir in the starter then start to warm the milk slowly, stirring as you do so. Be careful not to heat the milk too much; keep stirring until it reaches 37°C. Take regular readings in several places around the pan, rather than leaving the thermometer suspended in the milk in one place. Take it off the heat and let it rest for an hour.

Add the rennet and stir it in. Leave the milk to stand for 50 minutes or until the curds are set; they should still be soft.

Cut the curds into small pieces (see pp.52–3) and then use a hand whisk to break the curds up further so that they resemble cottage cheese curds. Transfer them to a sieve set over a large bowl to drain off the whey and reserve this to make ricotta (see p.102), whey butter (p.105) or a brine (see p.62).

When the curds feel a bit spongy, transfer them to a heatproof bowl and break them down with your fingers into small pieces, about 1cm cubed. Add a generous pinch of salt.

In a saucepan, bring 1 litre water to 95°C (almost boiling) and add it to the curds, stirring until they become melted into a single mass. Once cool enough to lift out with your hands, do so and shape into small balls by stretching and folding over. They should be smooth and springy. Leave to cool completely.

Serve straight away, or put the mozzarella balls into a container, cover with a little cold water and refrigerate. They will be good for 5 days.

Feta

It is possible to make feta using cow's milk but the fullest, most authentic flavour is achieved by using goat's or sheep's milk. Traditionally, the feta is submerged in a brine made of its own whey. The salty solution originally helped the cheese survive in the heat of the Greek climate but now it's really used to give it the classic sour, tangy flavours.

Before a cheese is placed in a brine it must be lightly salted all over and left to dry naturally at room temperature for 24 hours. This initial salting and drying helps the surface of the cheese become firm and prevents the cheese from breaking up when placed in the brine. Feta, or any brine-immersed cheese, will not develop moulds or significant rind on the surface because it is not exposed to the air.

A feta cheese can be ready to eat after just a week but there is really no limit to how long it can be left in the brine, as long as it is kept cool. The longer the feta is left in the brine the deeper the flavour will be.

Makes about 400g

4 litres whole cow's, goat's or sheep's milk (unhomogenised)
60ml active kefir starter (see p.72) or 1 unit of homofermentative mesophilic starter CHR-Hansen R-704 (see p.70)
¼ tsp liquid rennet (or 5 drops from a pipette) or ¼ dissolved rennet tablet
Sea salt

Special equipment

1 soft-cheese mould, about 16 x 8cm, or cheese press with follower

Put the milk into a preserving pan and slowly warm it to 32°C. Take regular readings in several places around the pan, rather than leaving the thermometer suspended in the milk in one place.

Leaving the pan on the low heat, add the active kefir or commercial starter. Using a pH meter, check the acidity at this point and once there is a lowering of the pH to 4.6 (i.e. higher acidity) remove the pan from the heat and cover with a clean tea towel. Leave to stand off the heat for 1 hour to allow the bacteria to culture and ripen the milk.

Now add the rennet and stir it gently into the milk. Continue to stir for a minute or two and then re-cover the pan with the tea towel to maintain the residual heat. The curds should show signs of setting after 20 minutes but they should be left for an hour after the rennet is added.

Cut the curds with a long-bladed knife (see pp.52–3). Gently stir the cut curds with your bare hands, which encourages the whey to come away so the curds start to become firmer. Pour off the whey into a measuring jug, making sure you have at least 1 litre for the brine.

Place the cut curds in a colander over a large bowl or the sink and allow the excess whey to drain off. Add a couple of generous pinches of salt to the curds; this draws more moisture out of them and will stop the curds from breaking up in the brine. Stir gently with your hands in the colander, then leave for a further 10 minutes.

If using a mould, place it on a draining mat or rack set over a large container. Pack the curds into the mould and leave to drain for about 3 hours or until reduced to a third of the original depth, then turn every 3 hours over a 12-hour period.

(Alternatively, you can place the curds in a press with a weighted follower, which will condense the curds much more quickly and give the classic, oblong feta shape. Remove, flip and re-press the curds every 10 minutes or so until they have cooled – the curds will be less susceptible to pressing once cooled.)

Remove the curds from the mould or press and leave them overnight on a draining rack covered with a loose sheet of muslin or cheesecloth.

The next day, make the brine: calculate 3% salt to the amount of whey; i.e. 1 litre whey will require 30g salt. Add the salt to the whey and stir until fully dissolved. You can make a stronger brine if you like (up to 10% salt) to significantly limit the bacterial growth, which will help preserve the cheese longer, though of course it will make it taste saltier too. Saltiness is a characteristic of feta, but at 10% salinity, the brine may be at the top end of acceptance for some. Pour the brine into a clean plastic, non-reactive bowl or tub, or into sterilised preserving jars (see p.40).

Salt the surface of the curds and leave to dry in a cool place for 24 hours (if your brining vessel is quite small, cut the feta into cubes before salting).

Now place the feta in the brine. Keep in a cool place for at least 2 weeks; the best results come from ageing for a further 2 months. The brine will work best below 10°C; i.e in a cool room or the fridge. The cooler the temperature the longer it will take for the feta to age. Some brines release gas when the cheese is submerged in them so, if you are using a jar with a lid, open it every so often to release any gas.

If you have left the feta in a block (rather than cut into smaller cubes), then lift it out of the brine and cut off what you need before replacing it in the brine. Of course, once you start eating your feta it will be so delicious that opening and closing the lid will become a regular occurrence. I always make a fresh brine each time I make a new cheese.

Ricotta

Ricotta is an Italian cheese traditionally made with the leftover whey from mozzarella, but you can use whey left over from any cheese-making. Making a whey cheese differs from making a curd cheese: the protein found in whey is albumin, which is not affected by rennet and does not coagulate at low temperatures, but it does form curds at higher temperatures. (*Ricotta* translates as 'cooked again' and it's called that because this will be the second time the liquid is heated, having started out as milk.)

Whey that is very acidic to begin with will not produce enough curds to form anything of substance. The whey left from making most renneted cheeses is considered 'sweet', whereas the whey from yoghurt cheeses or especially long fermented cheeses is considered 'acid'. If there is any doubt as to the acidity of your whey, measure it – if it registers a pH of 6 or higher, then it will be fine.

Ricotta is not a high yielding cheese – only a fifth of the total proteins in milk are whey proteins – but you can increase the yield by adding a small amount of milk or cream to the whey before heating (double cream gives a richer ricotta).

Makes about 400g

4 litres 'sweet' whey (see above)
400ml whole cow's milk or cream (single or double)
Juice of 1 lemon
A pinch of sea salt (optional)

Pour the whey and milk or cream into a preserving pan and heat slowly over a moderate heat. When the liquid reaches 85°C, add the lemon juice and allow the rolling bubble of the liquid to naturally mix it in – there is no need to stir it.

Remove the pan from the heat and allow to stand and cool slightly for 5 minutes. The ricotta curds will start to form shortly after and you may notice the whey become clearer or slightly yellow in hue.

Line a colander with muslin or cheesecloth (or use a ricotta basket mould, as shown, if you have one). Using a slotted spoon, remove the curds, which will be small and fluffy, from the whey and put them into the muslin-lined colander (or basket mould) to drain. Most of the whey will drain out in the first 5 minutes.

Once the ricotta reaches the desired texture (about an hour for a soft ricotta), you can use it or transfer it to a clean jar and store in the fridge, where it will keep for a week. If you stir in a pinch of salt, the ricotta will dry out and firm up and can be left for up to 4 weeks. This firmer ricotta can be grated or crumbled over food.

Whey butter

Whey is a by-product of the cheese-making process that can be used for making brine (see p.62), ricotta (see p.102) or this excellent butter. Made from the residual cream left in a batch of whey after the curds have formed, whey butter is more reminiscent of cheese than normal butter is, and it has a stronger flavour and oilier texture. Technically it isn't a cheese, but it feels as though it belongs here.

The amount of whey cream you have will depend on the freshness of the original milk. Whole milk that you buy (which is typically 3–4 days old) will have a layer of cream that has risen to the top and this will pass through to the whey, making it creamier.

Makes about 30g

300ml whey (minimum)
Sea salt

Leave the whey to stand at room temperature for 20 minutes then skim the cream from it or pass the whey through a fine muslin or cheesecloth so that the cream and liquid whey are separated; discard the liquid.

Put the whey cream into a food processor and blitz on a medium whisking speed until bubbles start to appear in the cream. At this point, switch to full speed and continue to blitz until the cream solidifies into butter.

Add a good pinch of salt and mould the butter into a round. Use immediately or wrap in baking parchment, refrigerate and use within 3–4 days.

Matured Cheeses

Anticipating the fruition of a matured cheese is one of my favourite parts of the process – whether it is waiting for the first signs of a white powdery bloom to appear on the surface of a Brie or watching for the orangey colour to develop on a pungent washed-rind cheese. When eventually you get to slice into a cheese that you have made and aged, the satisfaction of the complex, bold and nutty flavours makes all the waiting worthwhile. There's a whole range of different cheeses to try your hand at.

Soft mould-ripened cheese

The main characteristic of a soft mould-ripened cheese is the white mould rind that covers the entire cheese. They can be made from either cow's milk (such as the lovely Sharpham's Brie from Devon) or goat's (like the excellent Little Wallop from Somerset). The flavours range from creamy to nutty, and the textures are soft to the point of being runny. This category includes classics such as French Brie and Vacherin – though for the home cheese-maker, it makes sense to attempt smaller, more compact cheeses to start with.

These cheeses ripen from the outside in: the microflora which began as yeast in the original starter culture develop first on the rind. Supported by some salting, positive bacteria then permeate the paste of the cheese. The rind is the only part of the cheese that firms up, containing the oozing interior when it's fully ripe. Soft mould-ripened cheeses reach their peak at a few weeks of age.

Semi-soft, washed-rind cheese

Cheeses in this category have a pungent aroma. The rinds are initially washed in a brine solution, made with whey or alcohol in the form of beer, cider or spirits. The acidity in the wash helps to promote good bacteria, which produce sulphur compounds as they grow – hence the distinctive smell and robust flavour of the rind. These bacteria are also responsible for a colour change – the rinds usually have an orangey hue.

The paste of washed-rind cheeses is creamy. It may have a little resistance and elasticity, but will be nowhere near the firmness of, say, a Cheddar. Stinking Bishop and Ogleshield are my favourite washed-rind cheeses in the UK. Ogleshield is made from Jersey milk and washed in a special brine solution, while Stinking Bishop is washed in a brine made with perry.

Hard cheese

The curds for hard cheeses are often reheated or cooked after they have been cut to release more whey and create a firm texture; Parmesan is a good example. Alternatively, the curd may be milled or 'cheddared' into very small pieces to facilitate more whey drainage before moulding. The more compact and heavy the

press used to mould the cheese, the firmer the texture of the finished cheese. A hard cheese loses more moisture than other cheeses during its lengthy maturing. Fine examples of matured, hard cheeses include Barber's vintage Cheddar and Montgomery Cheddar.

Occasionally, particularly in hard cheeses that have been extra-matured, you will come across tiny, crunchy crystals. This is the effect of residual calcium combining with lactose to form 'crunchies'. Some consider this to be a defect in a cheese but I like it.

Alpine cheese This is a sub-category of hard cheese, traditional to a collection of farms in the Alps. The breeds of cattle used to make Alpine cheese yield milk with a high fat content, which reflects the wonderful pasture on which they graze over spring and summer. The cheeses are often studded with large holes: this is the effect of certain bacteria causing bubbles in the paste as the cheese ages. While most cheeses are aged at around 10°C, Alpine cheeses such as Emmental may be matured in a warmer environment, up to 24°C.

Blue cheese

Blue cheeses have either an interior veining of blue mould, or blue mould on the rind. They can be quite firm in texture (such as Stilton or Blacksticks Blue), or softer (like Cambozola). The mould develops from various strains of *Penicillium roqueforti*, which are added as part of the starter culture, and it eventually blooms on the cheese once it comes into contact with the air.

A process called 'needling' is responsible for the growth of the mould within the paste of blue-veined cheeses – a fine skewer is used to create a network of tiny tunnels, which allows contact with the air and encourages the development of interior mould.

Stilton is one of the most famous firm, blue-veined cheeses, yet it has altered its course away from its origins. It carries a Protected Designation of Origin (PDO) and must be made with milk from the counties of Derbyshire, Nottinghamshire or Leicestershire. This means, ironically, that 'Stilton' made in the original location from which it gets its name – which is in Cambridgeshire – cannot be called Stilton. There are some excellent 'Stilton-style' cheeses that are representative of the traditional methods (also excluded by the Stilton PDO), such as Dorset Blue Vinny, Colston Basset and Stichelton, which is made using raw milk.

Perhaps the most famous blue of all is Roquefort – a ripe, sheep's milk cheese made exclusively in the town of Roquefort in the south of France. The *Penicillium roqueforti* spores that create the distinctive blue veins can, legally, only be harvested in the Combalou caves near the town where they appear naturally. Commercially, *roqueforti* spores are available in dried culture form.

Brie-style cheese

This method will allow you to create a classic, soft, white, mould-ripened Brie- (or Camembert-) style cheese with a deliciously mushroomy flavour. It will also give you a good understanding of the 'full' cheese process and, once mastered, can be the start of all sorts of cheese adventures. You will need to allow a couple of weeks or so for the cheeses to mature before you can eat them.

Makes 3 small cheeses

9 litres whole cow's milk (unhomogenised)
1 unit of heterofermentative mesophilic starter (see p.70)
0.02 dose of *Geotrichum candidum* yeast
0.02 dose of *Penicillium candidum*
1.8ml liquid rennet (measured in a syringe for accuracy)
15g sea salt

Special equipment

3 soft-cheese moulds, about 10cm in diameter

Pour the milk into a preserving pan and warm it gently to 32°C. Take regular readings in several places around the pan, rather than leaving the thermometer suspended in the milk in one place.

When the milk reaches the correct temperature, remove from the heat and add the starter and the *candidum* ripening cultures, stirring them in thoroughly with a wooden spoon. Place a lid on the pan and leave the milk to stand for an hour to allow the starter to culture and ripen the milk.

Stir in the rennet and then leave undisturbed for another hour to coagulate. Check to see if flocculation has occurred (see p.51).

When you think the curds are set enough to cut, do the split test (see p.51). When they're ready, cut using the grid pattern (see pp.52–3). Stir the curds by hand – this is the best way of checking that the pieces are evenly sized; it's also pleasurable! The curds will start to sink to the bottom of the pan after stirring.

Stir regularly over a period of 30 minutes, or until all the curds have sunk.

Place the moulds on a draining mat set over a large container to catch the whey. Pour off about half the whey from the pan and then carefully ladle the curds into the moulds. Fill the moulds as much as you can with curd, then leave them for an hour at room temperature. The small amount of whey still in the curds will drain out of the moulds.

After an hour or so, the draining curds will have started to condense and will now fill only half the moulds. At this point they must be turned, which needs a bit of dexterity. Turn the round of curds out of the mould into your hand with a flip so that it is now upside down (see pp.54–5). Neatly place the round of curds back into the mould so that the side that was originally on top is now on the underside. Turn the cheese again after another 2 hours, and then again after a further 4 hours.

Remove the cheeses from the moulds after a total of 24 hours. You now need to rub a dusting of salt over the entire surface of each cheese. If you want to be really precise with this, then weigh the cheese and calculate 3% salt to the weight of the cheese before rubbing it on.

Put the cheeses on to a cutting board in an airy room and leave for around 6 hours, until the surface of each cheese starts to dry. Turn the cheeses a couple of times during this period.

When the surface is quite dry, place a large plastic box over the cheeses, which will create a microclimate with the correct humidity, and move them to a warm place, about 15–20°C. After 24 hours the rinds will start to show a slight furry quality and you will see the beginning of the white bloom.

When the first signs of bloom appear, move the cheeses to the fridge. The rinds will develop a full covering of white chalky mould after a couple of weeks. At this point the cheeses can be eaten, or wrapped in baking parchment to stop them from drying out too much and matured further. They will be good to eat up to 10 weeks from when they were first made.

Washed-rind cheese

Such cheeses are amongst my favourites because they carry pronounced flavours and aromas. As easy to make as Brie-style cheeses, their rinds are rubbed regularly with brine made from whey or alcohol as part of the ripening process. You'll need to allow around 5–7 weeks for ageing this particular brine-washed cheese.

Makes 3 small cheeses

4 litres whole cow's or goat's milk (unhomogenised)
Sea salt

plus
½ tsp liquid rennet (or 10 drops from a pipette)

For the starter

25g mesophilic starter MA-4001 (see p.70)
1g *Debaryomyces hansenii* yeast (DH Danisco)
1g *Geotrichum candidum* yeast (GEO17)
1g *Brevibacterium linens* (BL1)
1g *Penicillium candidum* (PC42 'VS')
or
60ml active kefir starter (see p.72)

For the brine

15g sea salt
1 litre whey, beer, cider or wine (see pp.62–3)

Special equipment

3 soft-cheese moulds, about 10cm in diameter

Pour the milk into a preserving pan and slowly heat it to a temperature of 32°C. Add the starter ingredients (or the kefir starter), stirring gently. Put a lid on the pan and remove from the heat so that it can be kept at 32°C while the starter cultures and ripens the milk. Leave for 1 hour.

Check the temperature of the milk: it needs to be at 32°C. Place the pan back on the heat to warm up again if necessary. Take off the heat, add the rennet and stir. Leave the milk to coagulate for an hour or so, placing a lid on top to keep it at 32°C. Check to see if flocculation has occurred (see p.51).

When you think the curds are set enough to cut, do the split test (see p.51). If it produces a clean break, cut the curds into 2cm cubes (see pp.52–3). At this point check the pH to see that it is moving down (towards 4.6).

Once the curds are cut, stir them with your hands before pouring off the whey into a measuring jug, making sure you get at least 1 litre.

Remove the curds from the pan by hand and pack them into the cheese moulds, filling the moulds as full as possible. Place these on a draining mat set over a large container to catch the whey.

After a couple of hours, the draining curds will have started to condense and will now fill only half the moulds. At this point they need to be turned. Tip the round of curds out of the mould into your hand with a flip so that it is now upside down (see pp.54–5). Neatly place the round of curds back into the mould so that the side that was originally on top is now on the underside. Leave to drain for a further 22 hours.

After a total of 24 hours, the cheeses will be firm enough to remove from their moulds. You now need to salt them lightly by rubbing a dusting of salt over the entire surface of each cheese. If you want to be precise with this, weigh the cheese and calculate 3% salt to the weight of the cheese before rubbing it on.

Place the cheese in your maturation 'cave' (see p.58) to begin ageing.

Once the cheese rinds feel dry, it is time to start brine-washing, so make a brine by dissolving the salt in your chosen liquid.

Wash the rinds using a muslin or cheesecloth soaked in the brine every other day, and flip them over each time so they can mature evenly. After 3–4 weeks the rind will show signs of the orange bloom and at this point you can stop washing with the brine. Allow the cheese to develop further pigmentation and aroma over the next 2–3 weeks before eating. It will be good to eat for another 3–4 weeks.

Caerphilly-style hard cheese

It takes a lot of commitment from an amateur cheese-maker to successfully make and age a slow-maturing hard cheese such as a Cheddar. I would never want to discourage anyone from this endeavour but, realistically, most of us will leave that kind of cheese to the professionals. However, there are some very good hard cheeses that don't require months of ageing and ripening. Caerphilly is a great example – it is made using the same technique as Cheddar but ripens in just a couple of weeks.

Makes 1 large cheese (about 1kg)

12 litres whole cow's milk (unhomogenised)

For the starter

1.5 units of homofermentative mesophilic starter CHR-Hansen R-704 or Danisco MA-11 (see p.70)

or

180ml active kefir starter (see p.72)

plus

3ml liquid rennet (measured in a syringe for accuracy)

12g sea salt

For the brine bath

1kg salt

4 litres water

Special equipment

A large hard-cheese mould (1.2kg) with follower

Gently heat the milk in a large stockpot until it reaches a temperature of 32°C. Remove from the heat and then add the starter ingredients (or the kefir starter), stirring gently. Put a lid on the pan and leave to stand for 2 hours to allow the starter to culture and ripen the milk.

Stir in the rennet and then leave the milk to coagulate for an hour. Check to see if flocculation has occurred (see p.51).

When you think the curds are set enough to cut, do the split test (see p.51). If it produces a clean break, cut the curds into 2cm cubes (see pp.52–3).

Return the pan to the heat and stir the curds with a wooden spoon over a period of 30 minutes, gradually increasing the temperature to 34°C. The curds will visibly shrink as they lose moisture during this time.

Remove the pan from the heat and stop stirring. Leave the curds in the whey until the pH measures 6.2. Carefully pour off the whey. Move the curds to one side of

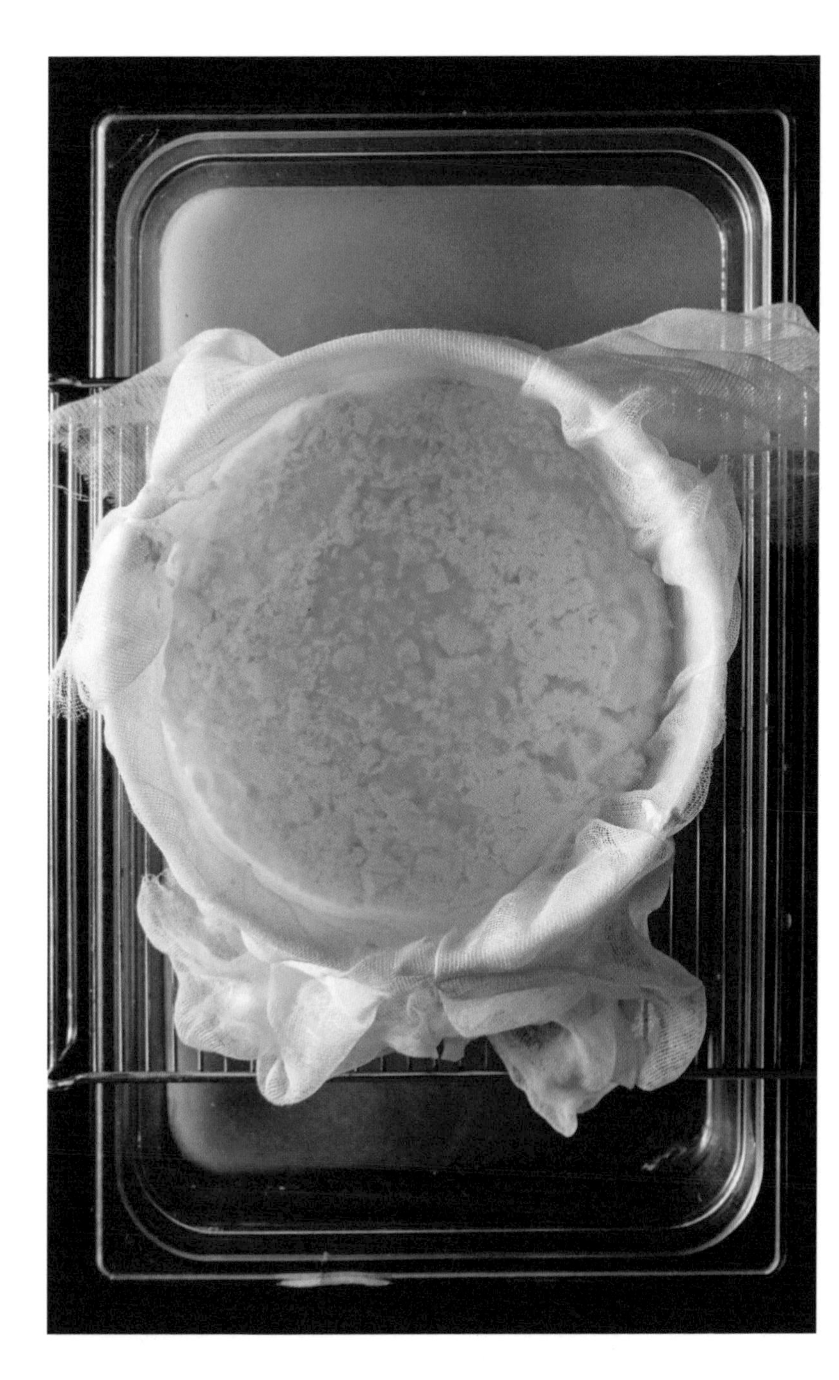

the pan and prop the pan up slightly on that side to encourage more whey to drain away from the curds. Check this whey with your pH meter and, when it has reached pH 6, pour it off.

Using a knife, cut the curds into individual small strips, about 2cm wide. Check the acidity of the curds using a pH meter. When the pH has lowered to 5.6, cut the strips into cubes. Sprinkle the salt over the chopped curd cubes. The temperature of the curds will have dropped but try not to let it get too much below 22°C or the curds might not bind together well.

Put a cheesecloth or muslin in the large cheese mould and then pack the curds into it. Lay the follower on top and then place a large lidded tub filled with water on top of that to weigh it down. Leave for 30 minutes.

Remove the weight and then turn the cheese over in the mould; the cloth will make this a bit easier. Reduce the amount of water in the tub by half, then place it back on top of the follower and keep in place for 12 hours.

Make a brine bath by dissolving the salt in the water in a large bowl. Turn out the pressed cheese into the brine so that it is fully submerged and leave for 6 hours.

Remove the cheese from the brine bath and place on a draining mat or rack set over a container for 24 hours.

Mature the cheese on a wooden board in a cold room (at around 10°C) or in your designated maturation 'cave' (see p.58), turning it every other day, for 2 weeks. It will be mature and ripe enough to eat at this point; it may also have developed a surface mould. The cheese will be good to eat for another 4–6 weeks.

Surface-ripened blue cheese

This is one of the easier, quicker blue cheeses. It is made using the same method as a Brie-style cheese, but the milk is inoculated with *Penicillium roqueforti*. The mould is limited to the surface rind of the cheese and does not travel into the paste. This is a great first blue cheese to make, as it ripens relatively quickly and is fairly small. Depending on your confidence and spirit of adventure, you might try using home-grown *Penicillium roqueforti* rather than buying it (see p.132). You'll need to allow 6–8 weeks for the cheese to age and develop its full flavour.

Makes 2 small cheeses (about 300g each)

9 litres whole cow's milk (unhomogenised)

For the starter

1 unit of heterofermentative mesophilic starter (see p.70)

0.02 dose of *Geotrichum candidum* yeast

0.02 dose of *Penicillium candidum*

or

125ml active kefir starter (see p.72)

0.02 dose of *Penicillium roqueforti* Danisco PRB6

or

120ml home-made *Penicillium roqueforti* solution (see p.132)

plus

1.8ml liquid rennet (measured in a syringe for accuracy)

2 tbsp sea salt

Special equipment

2 soft-cheese moulds, about 10cm in diameter

Pour the milk into a preserving pan and slowly heat it to a temperature of 32°C. Remove the pan from the heat and add your chosen starter ingredients. Stir them in, put a lid on the pan and then leave to stand in a warm place for an hour to allow the starter to culture and ripen the milk.

Add the rennet and stir it in gently. The curds will begin to set in about 10 minutes but should be left for another hour in the same warm spot.

When you think the curds are set enough to cut, do the split test (see p.51). If it produces a clean break, cut them using the technique described on pp.52–3. Stir the cubed curds gently by hand so that they begin to sink to the bottom of the pan. Do this 4 or 5 times over a period of half an hour.

Check the acidity at this point: the pH should be creeping down towards 5. Pour off most of the whey, making sure that you don't lose any of the curds.

Position your soft-cheese moulds on a draining mat or rack over a large container. The curds will still be very fragile, so use a ladle to transfer them to the moulds. Leave for an hour, by which time the curds should have compacted enough to be turned. If you are not confident of your flipping and turning technique (see pp.54–5), turn the curds into a spare mould, if you have one, rather than flipping it and returning it to the same mould. Repeat this after a couple of hours and once again after a further 5 hours; this will help with drainage.

The cheeses should be ready to take out of the moulds after draining for 24 hours in total. Check to see if the pH has reached the desired level of 4.6.

At this point, you need to salt the cheeses lightly by rubbing a dusting of salt over the entire surface of each cheese. If you want to be precise with this, then weigh the cheese and calculate 3% salt to the weight of the cheese before rubbing it on.

Place the cheeses on a board in an airy, warm room and leave until the rinds are dry, which can take anywhere between 6 and 12 hours.

Place the cheeses in your maturation 'cave' set-up (see p.58) to mature and ripen. Once or twice during the first week you should turn the cheeses over so that they don't stick to the board. After a week or so you should be able to see mould developing on the surface.

Once the cheeses have a good covering of mould they can be transferred to your fridge to finish off ripening. They will be ready after 6–8 weeks of ageing and good to eat for a further 4–6 weeks after this.

Gorgonzola-style blue cheese

This style of blue cheese is generally made using a large quantity of milk to produce a high volume of curds. The curds are salted before they fully form, which creates cracks and crevices within the cheese. The cheese is also skewered (or 'needled') to encourage the development of blue *Penicillium roqueforti* mould within the paste.

The true Gorgonzolas from the Piedmont and Lombardy regions of Italy are technically challenging to produce, involving making two batches of curds and draining one lot before adding them to fresher curds made the following day. They also use starter cultures at the bottom end of the thermophilic temperature range, which is difficult to replicate at home because the lower temperatures mean there is an increased risk of pathogens developing.

The recipe below will give you a Gorgonzola-style blue-veined soft cheese without the difficulty and risk. It follows the same method as the surface-ripened blue cheese on p.125, but calls for a stockpot to contain all 12 litres of milk. The cheese takes 2–3 months to age; the result is delicious.

Makes 1 large cheese (about 1kg)

12 litres whole cow's milk (unhomogenised)

For the starter

0.3 units of thermophilic starter STB-01 (Hansen) or equivalent (see p.70)

0.7 units of heterofermentative mesophilic starter Flora Danica (Hansen) or equivalent (see p.70)

0.02 dose *Debaryomyces hansenii* yeast (DH Danisco)

or

180ml active kefir starter (see p.72)

0.02 dose of *Penicillium roqueforti* Danisco PRB6

or

180ml home-made *Penicillium roqueforti* solution (see p.132)

plus

2.9ml liquid rennet (measured in a syringe for accuracy)

30g sea salt

For the brine solution

500g salt

2 litres water

Special equipment

A large-hard cheese mould (1.2kg) with follower

Pour the milk into a large stockpot and slowly heat it to a temperature of 32°C. Remove from the heat and add your chosen starter ingredients. Stir them in, put a lid on the pan and then leave to stand in a warm place for an hour to allow the starter to culture and ripen the milk.

Add the rennet and stir it in gently. The curds will begin to set after 10 minutes or so but should be left for another hour in the same warm spot.

When you think the curds are set enough to cut, do the split test (see p.51). If it produces a clean break, cut them using the technique described on pp.52–3. Stir the cubed curds gently by hand so that they begin to sink to the bottom of the pan. Do this 4 or 5 times over the period of half an hour.

Check the acidity at this point: the pH should be creeping down towards 5. Pour off most of the whey, making sure that you don't lose any of the curds.

Put the curds in a colander set over a big bowl and sprinkle the salt over the curds, gently stirring it in with your hands. This will stop the curds from knitting together too well. Allow to drain for a further 5 minutes.

Line the large mould with cheesecloth or muslin and place it on a draining mat or rack over a large container. Place the curds in the mould and leave in a warm room for 24 hours, flipping the curds 2 or 3 times during this time, using the cloth to help you as the curds are quite heavy.

The cheese will now be firm enough to remove from the mould. Lift it out and place it on a board. Leave, uncovered, to air-dry for 24 hours.

Place the cheese in your maturation 'cave' (see p.58) to ripen. This cheese likes very humid conditions and your cave should be at 90% humidity. Make a brine solution by dissolving the salt in the water in a large bowl. During the first week in the 'cave', wipe the cheese down with the brine solution at least twice to stop mould from forming on the rind. If it does, this isn't necessarily a bad thing – it's just not typical of a Gorgonzola-style cheese.

Once the cheese has been aged for a week, pierce it all over right though, from top to bottom and also through the sides, using a stainless-steel skewer. Place the cheese back in the 'cave' and turn it twice weekly to prevent it from sticking. The cheese will be ready to eat anywhere between 2 and 3 months, when it will be ripe and slightly soft to the touch. It will be good to eat for a further 4–6 weeks.

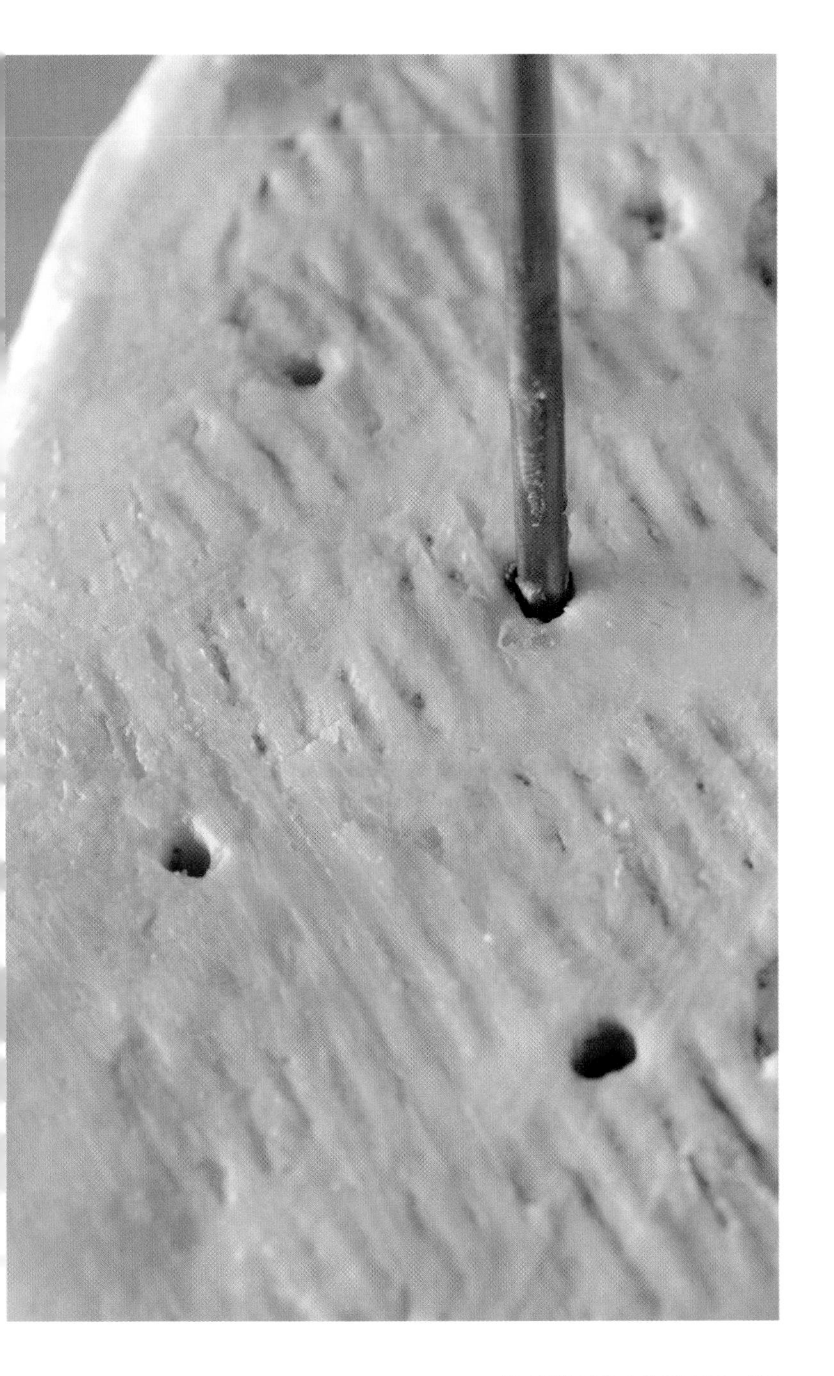

Penicillium roqueforti

I love making blue cheeses, despite (or perhaps because of) the challenges. The blue mould, *Penicillium roqueforti*, is a living organism that feeds off the fat and proteins in the cheese, if it is exposed to air. Sometimes salt is added to freshly made curds before they are moulded, to prevent them from knitting together too successfully, leaving natural air pockets in which the blue mould can develop. 'Needling' exposes the paste to even more air. Much like a mushroom, the spores develop mycelial roots that reach out through the cheese, creating those distinctive blue veins.

Harvesting *Penicillium roqueforti* yourself is extremely easy and there are two ways to go about it.

Method 1

Take a small pea-sized piece from a mature blue cheese containing a nice cluster of veins and crush it into 240ml water. Leave to stand at room temperature overnight. The water will then contain hundreds of thousands of spores that will thrive and develop. Strain just 60ml into a batch of 4.5 litres milk after it has been heated at the beginning of the cheese-making process to produce a good-sized blue cheese.

Method 2

For a particularly pure and strong culture, you can go back one step further and capture your very own blue mould by spreading a piece of sourdough bread with a layer of blue cheese (pic 1) and placing it in a sealed plastic food container at warm room temperature for 7–10 days. The mould will grow and start to cover the sourdough. Once the mould is firmly established, open the lid of the container to accelerate the drying out of the bread (pic 2). The acidity of the natural sourdough yeasts will stop any unwanted bacteria from forming alongside the blue mould. When you want to make the cheese, simply break off a few tiny pea-sized pieces of the mould-covered bread and drop them into 240ml water (pic 3). Strain the liquid through a muslin- or cheesecloth-lined sieve (pic 4), then use as described in Method 1.

1
2
3
4

Serving Cheese

If you store your cheese in the fridge, then you should take it out an hour or so before serving it, leaving the wrapper on. The flavour and texture will be at their best at cool room temperature. This is particularly true with a soft cheese: the paste should be creamy and slightly oozing from the rind but not losing its shape.

Slate or marble boards are commonly used to serve cheese on but, although they both help to maintain the temperature of the cheese and can make a real visual impact, I often use wood. I find cutting cheese on a hard, unforgiving surface slightly jarring and it can damage the knife. Wood is more yielding.

It is a good idea to designate one sharp knife to each cheese so they don't cross-contaminate flavours. For the real cheeseboard aficionado, a knife with a wide blade should be used for hard cheese, a narrow-bladed knife for semi-soft and a knife with a hollowed-out blade for soft cheese (so the cheese doesn't stick to it). You can also buy cheese knives with a double handle for cutting large cheeses, knives with fork-ended blades and boards that have cheese wires attached.

Cutting the nose off a wedge of cheese is considered a cheeseboard *faux pas*. The thin, pointed end (or apex) of a wedge is usually from the centre of the original wheel or cylinder, and therefore it's the ripest and probably best part of the cheese. It ought to be enjoyed by everyone. Cut a large wedge of cheese into smaller wedges, so each slice contains both the rind and the nose, and tells the whole story of that particular cheese. Similarly, with a very soft, well-ripened cheese, avoid scooping or 'mining' out the paste in the centre. This will leave you with a collapsing outer rind that tastes much better alongside the softer paste.

Most cheese rinds are edible, except those made from cloth or wax. Some people prefer not to eat the rind of soft cheeses – no one should feel bad about that but it's always worth giving it a go. Many soft cheese rinds have a gentle but comprehensive bloom of mould, which adds a mushroomy note to the cheese's overall flavour. Others are coated in ash, such as the wonderful Sleightlett goat's cheese. I always eat the rind of soft cheeses.

Brine-washed rinds offer some of the most diverse and complex flavours to be found in a cheese, and they sometimes have a fine covering of mould. Again, I always eat the rind because it has a wonderful texture and is all part of the cheese. The rinds of brine-washed cheeses do tend to release pungent odours, but they generally taste milder than they smell.

Hard cheese rinds are usually tougher than soft cheese rinds, and often have distinctly bitter flavour notes. I have no qualms about tucking into a Stilton crust because it has a pleasant, crumbly texture, and I do know of one person who is happy to chomp away at almost any hard cheese rind (the same person eats orange and lemon peels too), but you may prefer to eat up to the rind and then stop. There are excellent uses for cheese rinds; see my suggestions on p.179.

The cheeseboard

When it comes to eating your cheese, you'll appreciate it to the full if you serve it unadorned, just as it comes, as part of a cheeseboard or cheese platter. These cheese feasts traditionally end a meal, though they can also be a meal in themselves.

Good cheeses can be celebrated solo – there's no reason not to enjoy a single variety, such as a mature blue, ripened to perfection, after a meal. However, my favourite approach is to offer a selection of cheeses that show off a range of characteristics – perhaps one fresh cheese, one semi-soft, one brine-washed, one hard and a blue cheese, all together. It's nice to include cheeses made from more than one type of milk too. Ultimately, you must be guided by your own taste. For me, creating a cheeseboard is like picking your best fantasy football team from a squad of endless talent.

Accompaniments

Once you have settled on the combination of cheeses for your cheeseboard then it is worth spending some time choosing the accompaniments. The goal is to have cheese playing the starring role, with a supporting cast that brings out the very best in it. A few simple additions are all you need.

My preferred mode of cheese-to-mouth delivery is a cracker and I like to have several types available, ranging from neutral-tasting water biscuits through salty, peppery, herby or seedy crackers to plain oatcakes and even sweet digestives. I've yet to come across a cheese cracker combination I didn't like.

Bread is another classic cheese partner. However, while I'm all for bread-with-cheese as a snack or for lunch, I'm not sure bread has a place on the cheeseboard unless a really runny, soft cheese is included. If I'm putting together an antipasti board, however, with cheese accompanied by charcuterie, olives and dried tomatoes, I do add some olive-oil-soaked salty focaccia.

As well as crackers, there are other accompaniments to consider, such as fruit. (Fruit within cheese is an abomination of which we won't speak.) My favourite fruit and cheese pairing is pear with blue cheese – as long as the pear is ripe and juicy, but not drippingly so. Crisp apples are delicious with almost any hard cheese and other fruits including grapes and cherries can be lovely too.

The sharpness of a pickle is great for punctuating the rich, creamy tanginess of strong cheeses. Bullet-like cornichons and crisp, silver onions are my favourites, although I'm also partial to a good home-made piccalilli. A sweet quince jam goes particularly well with strong cheese. Soft cheeses, like ricotta, can be enhanced with a trickle of honey, or even a combination of sunflower seeds and walnuts toasted in a dry pan and tossed with a splash of soy sauce. Some chutneys are slightly overpowering on a cheeseboard and better suited to a ploughman's lunch.

Drinks to serve with cheese

What you choose to drink with your cheese is a matter of personal preference. There are no gastronomic rules, but here is a little advice to help you on your way.

Cheese and wine

Don't feel overwhelmed by the endless possible wine and cheese pairings. Rather, let your own instincts guide you – and look on the whole thing as a taste adventure. The idea is to find flavours that you think complement each other, without either the wine or the cheese being overpowered. In restaurants, cheese and wine from the same region are often served together. Even though there are many excellent wines produced here, this approach is a little too restrictive to be used as a general rule for British cheeses. So, consider the following tried-and-tested pairings.

- **Light- to medium-bodied dry white wines** are good with fresh cheeses, as well as goat's cheese.
- **Full-bodied dry white wines** go well with soft mould-ripened cheeses, such as Brie and Camembert.
- **Light and fruity dry red wines (and rosé)** also complement soft mould-ripened cheese but can be paired with soft blues too – like Cambozola or Birdwood Blue Heaven from Gloucestershire.
- **Medium-bodied dry red wines** go very nicely with rind-washed cheeses, such as Stinking Bishop.
- **Full-bodied dry red wines** pair well with strong, hard cheeses, such as Barber's Farmhouse Mature Cheddar.
- **Fortified wine** (notably Port) is generally paired with hard blue cheeses, such as Stilton.
- **Sweet dessert wines** are often served with cheese that is particularly salty, such as feta.

Cheese and beer

Over the last few years, the growth of the craft beer movement in this country has given us lots of new flavours and styles to choose from and the possibilities for pairing with cheese are endless.

- **Light beers** such as wheat beer or Pilsner lager match well with both fresh and soft cheeses, especially curd cheeses.

- **Malt beers and IPAs** are excellent with hard cheeses, such as good, nutty mature Cheddars.
- **Stouts and porters** go particularly well with blue cheeses, such as Blacksticks Blue.

Cheese and cider

Cider and cheese are often produced in the same regions – as with Cheddar and cider in Somerset, or Camembert and cider in Normandy. This is probably because good grazing pasture and apple orchards require similar topography and weather, particularly in terms of rainfall.

As with beer, British cider has moved up a gear or two and now ranges widely in style. You can still get the traditional flat, cloudy, scrumpy ciders but there's a whole range of sweet, bittersweet, bitter, dry and carbonated ciders too.

- **Carbonated ciders** are good with fresh and soft cheeses, such as Eldren, which is made in Dorset, or Oakdown from Devon.
- **Flat, tannic ciders** work well with hard sheep's cheese, such as Spenwood from Berkshire.
- **Tart ciders** are nice with goat's cheese and strong Cheddars, such as Somerset's Pennard Ridge.
- **Fruity ciders** go particularly well with brine-washed cheeses, such as Bude's Little Stinky.
- **Sweet ciders** go particularly well with blue cheese, such as Beenleigh Blue or Blissful Buffalo.

Cheese and non-alcoholic drinks

I was once served a wonderful cheeseboard of Shropshire Blue, Ogleshield and Montgomery Cheddar which came with three shot glasses containing freshly pressed apple, pear and grapefruit juice. At first, I thought the fruity shots were a gimmick but in fact they were delicious and made the cheeseboard as special as the cheese did. It makes sense when you think about it, as fruit is a great companion to cheese. This range of sweet, sharp and bitter juices both cleansed your palate and lifted the flavour of each cheese.

My favourite non-alcoholic partner for cheese is actually tomato juice: I find the richness of the juice works really well in this context. Experiment to find fruit juice and cheese combinations that you like. The only type of drink that I wouldn't ever pair with cheese is a hot one.

Recipes with Cheese

Cheese wafers

These crisp cheese delights are incredibly simple but they're so delicious it would be a crime not to include them. They are excellent on their own as a snack or canapé but can also be broken into salads or floated on a soup. I particularly like them with pears.

The beauty of this particular recipe is that it works with almost any semi-hard or hard cheese, whether it is mild, mature or blue-veined. The cheese is melted to form thin, crisp, savoury yellow snowflakes that crisp up as they cool.

Makes 10–15

250g semi-hard or hard cheese, finely grated

Heat the grill to high. Line a large baking tray with baking parchment. Using a tablespoon, place 10–15 little piles of grated cheese on the parchment, allowing plenty of room in between them for the cheese to spread.

Place the baking tray under the grill for 2–3 minutes until the cheese melts and spreads to form golden brown, lacy discs. Allow to cool and firm up on the tray for a few minutes before serving. The cheese wafers will keep for up to 3 days in a sealed container.

Cheese straws

Moreish and addictive, these cheese straws are the business. The warm, flaky rough puff pastry and savoury cheese are a perfect combination, and using equal parts lard, butter and cheese in the pastry makes it much lighter. Extra ingredients such as olives, sun-dried tomatoes or anchovies can add a layer of loveliness, and anyone who likes a bit of spice can pinch ½ tsp cayenne pepper into the mix along with the seasoning. Any hard cheese will be perfect for this snack, but my favourite is Lancashire Cheddar.

Makes about 12

100g plain flour, plus extra for dusting
50g cold butter, diced
50g cold lard, diced
50g Lancashire Cheddar or other hard cheese, grated
1 egg yolk
A splash of cold water
Sea salt and freshly ground black pepper

Sift the flour into a bowl and add the butter, lard and a pinch each of salt and pepper. Either in a food processor or by hand, rub these ingredients together until they resemble fine breadcrumbs.

Add the grated cheese and mix until just combined, then add the egg yolk and a splash of water and mix briefly until it comes together to form a stiff dough. Gather the dough and transfer it to a lightly floured surface. I like to roll and fold it as many as ten times so that it forms a stack of layers that can be distinguished from one another.

Flatten the dough into a disc, wrap in cling film and place in the fridge to rest for 20 minutes. Meanwhile, preheat the oven to 200°C/Gas mark 6 and lightly grease a baking tray.

Roll out the chilled pastry to a rectangle, about 5mm thick. Cut into strips, each about 10cm long and 1cm wide. Place these on the baking tray (it isn't necessary to twist them). Bake for 15–20 minutes until golden and puffed up.

If you have the willpower to allow your cheese straws to cool before eating them then you're a better person than I am. They would last for a week in a sealed plastic container, but that just isn't ever going to happen.

Cheese choux

These little buns may sound like inappropriate footwear, or a sneeze, but they're delicious. Flavoured with melting cheese and sprinkled with more cheese to serve, they make great snacks or canapés. Known as *aigrettes* in France, they are a bit fiddly but well worth the effort. The choux pastry can be prepared in advance, leaving just the frying in oil to finish off.

Makes 10–12

100g plain flour
½ tsp paprika
75g cold butter, diced
200ml water
3 medium eggs, beaten
75g Gruyère-style cheese or Appleby's Double Gloucester, coarsely grated
Vegetable oil for deep-frying
Sea salt and freshly ground black pepper
50g Parmesan or Old Winchester cheese, finely grated, to finish

Sift the flour, paprika and a good pinch each of salt and pepper into a bowl. Put the butter and water into a medium saucepan and heat gently. As soon as the butter melts and the liquid begins to boil, tip in the seasoned flour and immediately stir vigorously with a wooden spoon until the mixture forms a paste that comes away from the sides of the pan.

Remove the pan from the heat and allow the paste to cool for about 5 minutes, then gradually beat in the eggs, a little at a time, making sure each addition is fully incorporated before adding the next. The mixture should have a stiff dropping consistency and retain its shape on the spoon. Mix in the Gruyère-style or Double Gloucester cheese.

Heat about a 10cm depth of oil in a deep-fryer, or other deep, heavy pan suitable for deep-frying, to 180°C, or until a cube of bread dropped in turns golden brown in about a minute.

You will need to cook the choux buns in batches to avoid crowding the pan. Take a teaspoonful of the choux paste and drop it into the hot oil, using another clean spoon to help slide it off the teaspoon. Repeat to add another 7 or 8 spoonfuls and deep-fry for 3–4 minutes or until golden. Remove with a slotted spoon and drain on kitchen paper; keep warm in a low oven while you cook the rest.

Sprinkle the choux buns with the finely grated cheese and serve warm.

Broad bean crostini
with labneh

The combination of sweet young broad beans and lightly salted yoghurt curd is one of my go-to dishes when hosting one of our many outdoor meals with friends or family. Served on toasted slices of thin baguette, it makes a fresh and vibrant summer canapé or snack.

Makes 16

450g freshly podded broad beans (about 1kg in the pod)
60g hard sheep's cheese, such as Duddleswell, grated
Good-quality olive oil, to taste
½ lemon, for squeezing
16 small slices of ficelle or thin baguette, about 2cm thick
1 garlic clove, halved
250g labneh (see p.36)
Sea salt and freshly ground black pepper

Bring a pan of salted water to the boil and drop in the broad beans. Bring back to the boil and cook for just 2–3 minutes until tender. Drain the broad beans and run them under cold water to cool quickly, then pop the vibrant green beans out of their pale skins.

Put the broad beans into a bowl with the grated hard sheep's cheese and some salt and pepper. Using a fork or potato masher, mash the beans to a coarse paste, adding enough olive oil and lemon juice to achieve the texture and flavour you prefer. (Alternatively, you can use a food processor.) Taste and adjust the seasoning.

Heat the grill to high. Drizzle olive oil on one side of the bread slices and toast under the grill, then turn the slices and toast the other side. Rub the oiled side of the toasted bread with the cut garlic clove.

Spoon a generous dollop of the smashed broad bean paste on to each piece of toast. Tear or crumble the labneh over the top. Place on a serving board and trickle with olive oil, then give a final seasoning of salt and pepper. Eat right away… and wish you'd made more.

Saganaki cheese

I discovered this delicious warm appetiser on a recent family holiday to Crete. *Saganaki* means 'small frying pan' and relates to the little iron skillet used to cook the cheese. Often made from either goat's or sheep's milk, Greek cheeses are mostly aromatic, sweet and salty. Graviera – the cheese that I had in Crete – was made from a mixture of goat's and ewe's milk. Slightly off-white in colour, it has the texture of a semi-hard cheese like Gruyère. In traditional Greek restaurants, it is fried in the saganaki pan then doused in either raki or ouzo and set alight at the table to whoops of '*Opa!*'.

Serves 4–6

60g plain flour, for coating
200g slab of Greek-style saganaki cheese or a semi-soft goat's cheese, such as Posbury or Puddle
1 tbsp olive oil
1 shot of ouzo or raki

Spread the flour out on a plate. Place the cheese on the flour and turn it, to coat each surface in a dusting of the flour.

Heat the olive oil in a frying pan then add the cheese and sear on each side until golden brown, turning it with tongs. The cheese should have a light crust with a melting centre.

Take the pan off the heat and add a splash of ouzo or raki. Ignite with a long match at the table and muster the enthusiasm to shout '*Opa!*'. Or, failing that, knock the shot back in the secrecy of the kitchen.

Croque monsieur

This snack was the first French food I ever tried – while interrailing across Europe in the mid-1980s – and it reminds me of Parisian train stations. In reality, a croque (from *croquer* meaning 'to bite') monsieur is just a glorified ham and cheese toastie but the French find several ways in which to elevate it. Alternative versions include tomato (*croque provençal*), sausage (*croque andouille*), salmon (*croque norvégien*), sliced potatoes and reblochon (*croque tartiflette*) and, of course, *croque madame* which has a fried egg on top. The original cheese favoured in France was Emmental or Gruyère, so any of the Alpine cheeses would keep you on a traditional path – but you can veer off-piste with another cheese of your choice.

Serves 2

30g butter
4 slices of brioche or soft bread
1 tsp Dijon mustard
100g Alpine-style cheese or Ogleshield, grated
2 slices of thick-cut, traditionally cured ham
2 heaped tbsp Cheaty cheese sauce (see p.167)

For croque madame (optional)

2 eggs
A little rapeseed or olive oil

Preheat the oven to 180°C/Gas mark 4.

Melt the butter in a frying pan until it begins to foam. Add the slices of brioche and fry on both sides until lightly browned. Remove from the pan. (If you want to make croque madame, retain this pan for the eggs.)

Spread two of the browned brioche slices with an even layer of Dijon mustard, on one side only. Sprinkle half the cheese over the mustard and place a slice of ham on top. Scatter over the rest of the cheese, dividing it equally. Sandwich together with the remaining brioche slices.

Place a heaped tablespoonful of cheese sauce on top of each sandwich and spread it out so it reaches the edges. Transfer the sandwiches to a baking tray and place in the oven for 4–5 minutes until browned on top and melted in the middle.

If you want to make croques madame, while the sandwiches are in the oven, fry two eggs in a little oil and the residual butter in the frying pan.

Transfer the croques monsieur to plates, top with the fried eggs, if using, slice in half and enjoy a profound sense of *joie de vivre*!

Drunken Welsh rarebit

Welsh rarebit is a typical River Cottage snack, which my daughters often call 'Welsh rabbit'. This version is more for the grown-ups though, as it uses beer instead of milk to create a mellow, hoppy, creamy cheese sauce. The trick is to heat it all together slowly with a dollop of mustard, stirring continuously so that it thickens nicely and doesn't burn off too much of the beeriness. For an additional kick, I like to add cayenne pepper and chilli sauce, which gives the 'Welsh rabbit' a bit of dragon's fire.

Serves 2 or 4

130g Caerphilly, grated
2 tsp plain flour
2 tsp mustard
130ml beer
4 slices of thick-cut farmhouse bread
A pinch of cayenne pepper
A splash of chilli sauce

Stir the grated cheese, flour and mustard together in a saucepan over a low heat. Add the beer, stirring until the mixture thickens to a creamy texture. Do not allow it to boil.

Heat the grill to medium-high. Toast the slices of bread on both sides and then dollop the cheese and beer sauce on top, spreading it out to the edges. Place the slices of cheesy toast under the grill and cook until the cheese is browned.

Finish by sprinkling each slice with a pinch of cayenne and a splash of chilli sauce.

Scape goat cheese

After the rush of wild garlic in late spring come 'scapes' – long green stems, shooting out of the garlic bulbs hidden below. Left alone, these would flower in the summer, but because that diverts food from the garlic bulb they are better snipped off. Fortunately scapes are delicious to eat. Firm and crunchy, they are somewhere between asparagus and green beans in texture, with a mild garlicky taste. Scapes can be eaten raw, cooked briefly in boiling salted water, or steamed. I have paired steamed scapes with goat's cheese curds – not merely because I like the play on words, but because the combination is delicious.

Serves 2

6–10 garlic scapes
100g fresh goat's curd cheese (see p.95) or firmer goat's cheese, crumbled or grated
Sea salt and freshly ground black pepper
Extra virgin olive oil or clear honey, to finish (optional)

Steam the garlic scapes or immerse in boiling water for a couple of minutes to blanch them and then drain well.

While they are still warm, arrange the scapes on a plate and scatter the goat's curd or cheese over them. Season with salt and pepper to taste. An additional swirl of olive oil or honey on top turns the dish into a complete winner.

Fennel and apple salad
with goat's cheese

This is a refreshing salad that just needs a trickle of olive oil to finish it off. It can be served as a starter or as a light lunch, accompaniment or side dish. Its simplicity belies its versatility and the muted colours give the salad an understated elegance.

Serves 4

2 medium fennel bulbs
2 tbsp olive oil
Juice of 1 lemon
2 medium apples
100g hard goat's cheese, such as Quicke's
Sea salt and freshly ground black pepper

Trim the base and top of the fennel, reserving any feathery fronds, and discard any tough outer layers. Slice the fennel bulb as thinly as you can from top to bottom and place in a bowl. Add the olive oil and lemon juice, toss the fennel to coat and season with salt and pepper. Set aside to marinate for 30 minutes.

When you are ready to serve the salad, core and slice the apples and add them to the fennel. Toss to combine.

Divide the salad between individual plates and then shave the goat's cheese over the top. Check the seasoning and serve.

Pear and blue cheese salad

with figs and pomegranate

This is a bitter-sweet salad with flavours at both ends of the spectrum – you really need bold flavours to compete with the full, stinky qualities of blue cheese. A good, strong Gorgonzola-style cheese works well in this salad, or try a Blackstick's Blue from Butler's in Preston if you want that pleasing orange paste (which is achieved by mixing annatto into the cow's milk when the cheese is being made). Choose pears that are soft and ripe.

Serves 4

2 ripe dessert pears
Juice of 1 lemon
2 figs
1 pomegranate, halved
1 radicchio, trimmed and separated into leaves
200g full-flavoured blue cheese, crumbled

For the dressing

1½ tbsp red wine vinegar
1 tbsp wholegrain mustard
2 tbsp clear honey
2 tbsp olive oil
Sea salt and freshly ground black pepper

Quarter the pears lengthways and remove the core, then peel them if you prefer. Put the pears into a bowl, sprinkle with the lemon juice and toss to coat. Cut the figs into slim wedges and add to the bowl.

Hold each pomegranate half cut side down over a board and whack the skin side with a wooden spoon to release the seeds.

For the dressing, whisk together all the ingredients in a bowl, seasoning with salt and pepper to taste.

Pour the dressing into a salad bowl, add the radicchio leaves and toss them to coat. Divide the dressed leaves between individual plates and top with the pears and figs. Scatter over the blue cheese and pomegranate seeds, then serve.

Cheese and chard soufflés

The thought of making a soufflé may strike fear into the heart of any amateur domestic god or goddess, but this version is simple. It just needs a little preparation, and diners who will be ready to tuck in as soon as the soufflé is out of the oven. To get an even distribution of heat, I use a bain-mairie. Another trick is to line the buttered ramekins with fine, cheese-flavoured crumbs – this will encourage the soufflé mix to rise. It's possible to make these soufflés with any cheese; you can also use spinach instead of the chard.

Serves 8

600ml whole milk
1 bay leaf
1 shallot, peeled
2 cloves
100g butter, plus 10g for greasing
200g hard cheese, grated, plus 10g finely grated for the ramekins
10g fine breadcrumbs (sourdough is my preference)
100g plain flour
8 medium eggs, separated
1 tsp Dijon mustard
200g shredded, cooked and drained rainbow chard or spinach
A pinch of sea salt

Special equipment
8 ramekins (8cm diameter, 5cm deep)

Preheat the oven to 200°C/Gas mark 6.

Bring the milk to the boil in a saucepan and add the bay leaf and the shallot studded with the cloves. Remove from the heat and set aside to infuse.

While the milk is infusing, grease the ramekins with the 10g butter, making sure you cover the base and sides.

Combine the 10g finely grated cheese with the breadcrumbs and sprinkle the mixture over the base and sides of the ramekins so there is an even coating adhering to the butter.

Melt the 100g butter in a saucepan over a medium-low heat. Stir in the flour to create a paste (roux) and cook gently, stirring often, for 4–5 minutes (to lose the raw taste of the flour). Slowly add the strained, infused milk to the roux, whisking to avoid lumps. Cook gently, stirring frequently, until thickened.

Stir in the 200g grated cheese, then add the egg yolks, stirring all the time so that they give the sauce a glossy shine. Now stir in the mustard and the chard or spinach. Season with the pinch of salt and remove from the heat.

Whisk the egg whites in a large, clean bowl to soft peaks and then fold them into the soufflé mix, using a spatula. Spoon the mixture into the ramekins and level them off with a palette knife so that they are filled to the brim. Run the tip of your thumb or a clean cloth around the inner rim of each dish to create a small gap – this releases the mixture and encourages it to rise evenly.

Stand the ramekins in a deep roasting tray and pour boiling water into the tray to come halfway up the sides of the dishes. Place in the hot oven and bake for 10–12 minutes until the soufflés have risen. Avoid the temptation to open the oven door to check on them as this will lower the oven temperature and may cause your soufflés to collapse. Serve the soufflés straight away.

Baked Brie with truffle,
hazelnuts and thyme

The only thing better than a ripe Brie, barely containing its creamy paste within the thin, white-bloomed shell, is a ripe Brie that has been baked and adorned with crushed hazelnuts, shavings of truffle and sprigs of thyme. All the textures and tastes meet and create a rich alchemy that gives you a feeling of utter wellbeing. You could also try topping the Brie with sweet caramelised onions or sharp fruits such as red berries or cranberries. I prefer to keep the cheese in the round wooden box it came in for baking.

Serves 2

A round of Brie, 250–300g, such as Lubborn
A handful of sprigs of thyme
8–10 hazelnuts
A small truffle
A baguette, to serve

Preheat the oven to 200°C/Gas mark 6.

Remove any wrapping from the Brie and return it to its wooden container, with the lid off. (If your Brie doesn't have a box, you can buy a special-purpose Brie baking dish or use any ovenproof dish which will hold the cheese snugly.)

Scatter most of the thyme sprigs on top of the Brie and bake for about 10 minutes until the cheese is bubbling under the surface and slightly browning around the rim. Meanwhile, crush the hazelnuts, using a pestle and mortar.

Sprinkle the crushed hazelnuts over the hot Brie and shave the truffle thinly but generously on top, using a mandoline or a swivel vegetable peeler. Scatter over the remaining thyme sprigs and eat immediately, with a fresh, crusty baguette.

Cheaty fondue
or cheese sauce

I especially love dipping freshly steamed asparagus into this fondue, but it is equally good as a cheese sauce. The addition of double cream makes the texture thick, velvety and smooth. I prefer to use a strong, mature Cheddar to flavour it; I've experimented with many different types, and to this day have never been disappointed with the result.

It's traditional, when making a fondue sauce, to add a little dash of white wine. However, I'm such a fan of Black Cow Cheddar and the award-winning vodka distilled from the whey (which would otherwise be wasted) that I combine the two in this recipe – awarding myself an extra lick of the spoon for being clever with both food waste and flavour.

Serves 4–6

30g butter
30g plain flour
300ml whole milk
300ml double cream
250g Black Cow or other mature Cheddar, grated
35ml vodka (ideally Black Cow)
Sea salt and freshly ground black pepper

Melt the butter slowly in a heavy-based saucepan over a medium-low heat. Add the flour and stir until it is incorporated into a soft paste. Cook gently, stirring often, for 4–5 minutes (to lose the raw taste of the flour). Gradually pour in the milk, whisking well to avoid lumps. Lower the heat slightly and cook gently, stirring occasionally, for 10–15 minutes.

Stir in the cream and continue to heat gently for a further 5 minutes.

Remove from the heat, add the grated cheese and stir until it has melted. Add the vodka and season with salt and pepper to taste.

The sauce can be used immediately – for fondue the heat from a fondue set burner will keep it at a good temperature. In my house, anything goes when it comes to ingredients for dipping in the sauce.

Raclette

Raclette is both a type of cheese and a classic dish originating from Switzerland (and also popular in France). Large rounds of cheese are cut and heated so that the exposed face of the cheese melts and can be scraped off on to a diner's plate. Traditionally the raclette would have been heated over an open fire by Alpine herdsmen, but today it is more likely to be heated at the table using an electric grill and heated griddle plate. Authentically, raclette is melted over small, firm potatoes and accompanied by gherkins, pickled onions and cured meats. Of course, you can enjoy the cheese simply sliced and melted over potatoes under the grill – but in our family, the full-on version is a wild camping, open-fire favourite.

Serves 2–3

500g new or small waxy potatoes, such as Pink Fir Apple
300g raclette or other cheese with good melting qualities, such as Ogleshield (pre-sliced if using the grill)

For the dill cucumber pickle

1 small onion, peeled and sliced
1 tsp caster sugar
2 tbsp white wine vinegar
1 tsp chopped dill
1 small cucumber, thinly sliced
Sea salt and freshly ground black pepper

To serve

A selection of cured meats

First make the dill cucumber pickle. Spread the onion out in a shallow dish and pour on enough boiling water to cover. Leave to soak until the water is cold. Meanwhile, combine the sugar, wine vinegar and a little salt and pepper in a small pan and heat gently until the sugar is dissolved. Remove from the heat and stir in the chopped dill.

Drain off the water from the onion, then add the cucumber and pour on the hot vinegar mixture. Leave to macerate for at least 1 hour, or preferably overnight.

When you are ready to eat, boil the potatoes in their skins until tender, then drain well and slice in half.

Place the potatoes, cut side up, in a roasting tray, top with slices of cheese and grill until melted. Alternatively, place the open face of the cheese close to an open fire and scrape off the surface as it melts, on to the potatoes.

Serve immediately, with the dill cucumber pickle and a selection of cured meats.

Spanakopita

This traditional Greek pie is like a beautifully wrapped present – equally appealing before and after you have seen what's inside. It can be eaten hot or cold. The filo casing is light and crisp, and the filling is made with spinach and feta – which gives it a lovely saltiness. If you prefer less salt, either swap the feta for a less salty, crumbly cheese like Cheshire, or simply go easy on the seasoning.

Serves 6

300g washed and trimmed spinach
4 spring onions, trimmed and chopped
300g feta, such as Medita, or Cheshire, coarsely grated or crumbled
2 medium eggs, beaten
2 tsp chopped parsley
2 tsp chopped dill
8 sheets of filo pastry, each about 30 x 20cm
150ml olive oil
Sea salt and freshly ground black pepper

Preheat the oven to 180°C/Gas mark 4. Have ready a rectangular pie dish, about 20 x 15cm.

Bring a small amount of water (about a 3cm depth) to the boil in a large pan, add the spinach and blanch for 1 minute. Drain the spinach in a colander and run under a cold tap to refresh the leaves. Allow the spinach to drain thoroughly and then give it a good squeeze to remove all excess water.

Tip the spinach on to a board and chop it up, then place in a large bowl with the spring onions and cheese. Pour in the beaten eggs, add the chopped herbs and season with salt and pepper. Mix well.

Brush one sheet of filo pastry with olive oil and place it in the pie dish, leaving the edges overhanging the sides of the dish. Layer another 3 filo sheets on top, brushing each one with oil first and placing them at different angles, to create a pie casing.

Spoon in the feta and spinach mixture, then cover with all but one of the remaining filo sheets, brushing each with oil and positioning them at different angles, as before. Fold all of the overhanging filo back over the top. Brush the final filo sheet with oil, then scrunch and place it on top of the pie.

Brush the surface of the pie with more oil. With the tips of your fingers, sprinkle a little cold water over the pastry to stop excessive curling in the oven. Bake for 30–40 minutes, until the pastry topping is crisp and golden.

Chicken and leek lattice

Originally, this savoury plait was a warming Monday supper treat in our house, made with the leftovers from a Sunday chicken roast and richly flavoured with leek, mushrooms and Cheddar. As our children have grown, there is no longer such a thing as leftovers, so we now make a family-sized version using a whole medium, organic chicken. It's well worth it.

Serves 6

500g cold roast chicken (off the bone)
20g butter
1 large leek, washed and thinly sliced
2 garlic cloves, peeled and grated
100g mushrooms (ideally a mix of chestnut and wild, such as cep or chanterelle), cleaned and sliced
Finely grated zest of 1 lemon
1 tbsp chopped parsley
500g ready-made all-butter puff pastry (2 x 250g blocks)
200g mature Cheddar, such as Montgomery, grated
1 medium egg, beaten
Sea salt and freshly ground black pepper

Preheat the oven to 200°C/Gas mark 6. Shred the cooked chicken and set aside.

Heat the butter in a frying pan over a medium heat. Add the leek and garlic and sauté for a few minutes until softened. Add the mushrooms and cook for about 5 minutes. Take the pan off the heat and stir in the lemon zest and chopped parsley. Set aside to cool completely, then add the shredded chicken and stir to combine. Season with salt and pepper to taste.

Roll out each block of puff pastry on a lightly floured surface to a large rectangle, about 35 x 25cm. Sprinkle the cheese evenly over both sheets of pastry, then fold each pastry sheet in half, enclosing the cheese. Roll out again so that the cheese and pastry are integrated, to two 35 x 25cm rectangles.

Lay one pastry sheet on a non-stick baking tray with a longer side facing you, brush lightly with water and place the other pastry sheet on top. Spoon the filling evenly down the middle, leaving a 4cm margin at the top and bottom and a wider margin of about 10cm on both sides. From the sides, cut diagonal strips in the pastry towards the filling. Brush the strips with beaten egg and fold them alternately over the filling, roughly 2cm apart, so they overlap in the middle to form a simple plait enveloping the filling. Tuck the ends underneath.

Brush the plait with beaten egg and bake for 30 minutes or so, until the pastry is crisp and golden. Slide the plait on to a board and rest for 5 minutes before serving.

Cheese and onion tart

If there is a recipe that shouts more of the lost county of Lancashire than a cheese and onion tart, I will eat my proverbial flat cap – I would be discrediting my heritage if I did not include it! You have to resist the temptation to put bacon (or indeed potato) in this tart, and instead champion its simplicity. The addition of bacon would turn the tart into a quiche – which is a bit like calling Lancashire 'Greater Manchester'. My recipe is inspired by one of Lancashire's great chefs, Nigel Howarth, who helps to keep many of the region's ingredients on the map.

Serves 6–8

For the pastry

200g plain flour, plus extra for dusting
A pinch of sea salt
100g butter, diced
2 medium eggs, beaten

For the filling

25g butter
2 large onions, peeled and thinly sliced
3 medium eggs
300ml crème fraîche
¼ tsp freshly grated nutmeg
100g hard cheese (Lancashire of course!), grated
150g fresh goat's curd cheese **(see p.95)**
Sea salt and freshly ground white pepper

Special equipment

23cm loose-based tart tin

To make the pastry, sift the flour and salt into a bowl, then rub in the butter with your fingertips until it resembles fine crumbs. Add the beaten eggs and bring the dough together with your hands, adding a trickle of cold water if necessary. Knead the dough lightly until smooth and silky, then flatten to a disc, wrap in cling film and chill in the fridge for 20 minutes or so. Preheat the oven to 200°C/Gas mark 6.

Roll out the chilled dough on a lightly floured surface to about a 3mm thickness and use to line the tart tin; press the pastry into the corners and sides of the tin and make sure it extends above the rim by 5mm–1cm.

Line the pastry case with a sheet of baking parchment and add a layer of baking beans or dried pulses. Bake 'blind' for 15 minutes. Lift out the paper and beans and return the pastry case to the oven for about 5 minutes until the pastry is dry and lightly coloured. Remove from the oven and set aside.

Lower the oven setting to 180°C/Gas mark 4. Once the pastry has cooled a little, trim away the excess pastry from around the rim using a sharp knife.

To make the filling, melt the butter in a pan, add the sliced onions and sauté gently for 10–12 minutes until soft and lightly caramelised. In a bowl, mix the beaten eggs with the crème fraîche, nutmeg and some salt and white pepper.

Spoon the caramelised onions into the pastry case and scatter the grated cheese and goat's cheese over them. Carefully pour on the egg mixture. Bake in the oven for 35–40 minutes or until the filling is set, with just a pleasing wobble.

Transfer the tart, still in its tin, to a wire rack and leave to cool for about 15 minutes before easing it out of the tin and serving warm.

Broccoli and cheese soup

Blue cheese and broccoli is one of the best soup combinations, in terms of taste and aesthetic. Broccoli is particularly good at retaining its flavour, even when it has been through a blender, and the addition of a strong blue cheese matches it for robustness and texture. For an extra layer of cheesiness, you could adorn the soup with cheese wafers.

Serves 4

50g butter
1 onion, peeled and chopped
1 medium potato, peeled and chopped
500g purple sprouting broccoli, chopped
1 litre vegetable stock
250g blue cheese, such as Dorset Blue Vinny, grated
Sea salt and freshly ground black pepper
A drizzle of extra virgin olive oil and/or cheese wafers (see p.145), to serve

Melt the butter in a large pan over a low heat, then add the onion and potato and stir. Cover and cook gently for 15 minutes.

Add the broccoli, pour in the stock and bring to a simmer. Season with salt (lightly, as the cheese will add more) and pepper. Cook for a further 10 minutes or until all the vegetables are tender. Drain them in a sieve over a jug to catch the liquor.

Blitz the vegetables with a third of the liquor in a freestanding blender, or using a stick blender in the original pan. Gradually blend in the reserved liquor until the soup becomes silky and smooth.

Reheat gently in the pan to just below boiling. Take off the heat and stir in the blue cheese until melted and combined. Taste and adjust the seasoning as necessary.

Ladle the soup into warmed bowls and add a grinding of pepper. Top with a simple drizzle of olive oil and/or a cheese wafer to serve.

Cheese rind stock

The notion of 'leftover' cheese is an odd one in our house. If cheese is around, it generally gets wolfed down. I enjoy the musty crust of almost any cheese but there are some very hard rinds, such as Parmesan or Manchego, that even I can't manage. Thankfully, these make an excellent addition to soups, stews, casseroles or sauces. Strangely, the rinds don't add an element of cheesiness but instead an injection of the magical umami flavour which gives any dish extra oomph.

Add the rinds to a soup, stew, casserole or sauce about 20 minutes before the end of the cooking time, to allow time for them to infuse. The rinds don't render down and disappear but soften and become a bit gluey, so you may want to fish them out before serving.

P.S. I like to add cheese rinds to home-made stocks too. Another good use is to add them to boiling water before cooking pasta in it.

Cauliflower cheese

Cauliflower cheese is a family favourite in the Lamb household and once it's dished out, the race is on to be first to finish – either for second helpings or to fight over the hard, darkened cheesy scrapings burnt on to the rim of the baking dish. For such a simple recipe, there can be umpteen variations. My wife likes to steam the cauliflower florets and let them cool before coating in the sauce and baking in a ceramic dish. I prefer to pour the sauce over the hot cauliflower then grill it.

Serves 2 as a main dish, 4–6 as a side

500ml whole milk
1 small onion, peeled but left whole
1 bay leaf
100g butter
50g plain flour
100ml double cream
250g Cheshire cheese, grated
1 medium-large cauliflower
Sea salt and freshly ground black pepper

Pour the milk into a saucepan and add the onion and bay leaf. Heat gently to a simmer, then take the pan off the heat and set aside to infuse for 15 minutes. Remove the bay leaf and onion and season the milk with salt (lightly, as the cheese will add more) and pepper.

In another heavy-based saucepan, melt half of the butter over a medium heat and then stir in the flour to make a paste. Cook gently, stirring, for a couple of minutes to make sure that the flour is cooked.

Now slowly add the infused milk to this paste (called a 'roux'), whisking all the time over the heat to keep the mixture smooth. Add the cream and grated cheese, still whisking. When the cheese has fully melted and the sauce has thickened, remove the pan from the heat. Taste and adjust the seasoning if necessary.

Heat the grill to high and bring a large pot of salted water to the boil. Cut the cauliflower into florets, but try and keep as much stalk attached as possible. Add to the boiling water and cook for 3 minutes, then drain thoroughly.

Melt the remaining 50g butter in a frying pan, add the blanched cauliflower florets and cook over a medium heat until they start to colour. Transfer to an oven dish and pour the cheese sauce over them. Place under the grill and cook until the sauce is browned and bubbling.

Serve the cauliflower cheese straight away, either as an accompaniment to another dish or just on its own.

Cheese quesadillas

I love Mexican food – not least the savoury snacks made with warm, floury tortillas. This simple quesadilla is a filled tortilla, topped with cheese and folded, then warmed in a pan until the cheese is melted and gooey. There are no limits to the filling ingredients you can use but – as with pizzas – I think it is better to keep it simple and not to overload the tortilla. You don't need to use an authentic Mexican cheese, but do choose a cheese that has good melting qualities.

Serves 4

40g butter
4 large flour tortillas (23–25cm in diameter)
400g mature Cheddar or Cornish Gouda, grated

For the filling
400g tin plum tomatoes in juice
1 tbsp olive oil
1 small red onion, peeled and finely chopped
2 ripe avocados, peeled, stoned and chopped
Juice of 1 lime
A handful of coriander, roughly chopped
Sea salt and freshly ground black pepper

For the filling, roughly chop the tinned tomatoes and tip them into a large bowl with their juice. Add the olive oil, red onion, avocados, lime juice and coriander and mix gently to combine.

Warm a 28–30cm frying pan over a moderate heat and add a knob of the butter. Once it has melted, place a tortilla in the pan and wait until it starts to crisp. Take roughly a quarter of the grated cheese and sprinkle it on top of the tortilla while it is still crisping. When the cheese begins to melt, spoon roughly a quarter of the filling over one half of the cheese (this will make it easier to fold over the tortilla).

After a minute or so, fold the tortilla over to enclose the filling; the underside should be dotted with brown spots. Keep warm in a low oven while you repeat the process with the remaining tortillas and filling, adding more butter to the pan as necessary. Transfer the tortillas to a board and cut them in half to serve.

Macaroni cheese

One of the most popular comfort foods, macaroni cheese is all about simplicity, familiarity and cheesy indulgence, and nothing else will suffice when you have the craving. It can be loaded with one or two extra ingredients but really it's best kept simple. This is the classic version that should satisfy those cheesy, creamy desires.

Serves 2

200g dried macaroni
35g butter
25g plain flour
450ml whole milk
½ tsp English mustard
Freshly grated nutmeg
50g mature Cheddar, grated
50g Parmesan or other flavourful hard cheese, such as Old Winchester, grated
Sea salt and freshly ground black pepper

Bring a large pan of salted water to the boil, add the macaroni and cook until *al dente* (tender but firm to the bite).

Meanwhile, melt the butter in a pan, stir in the flour and cook, stirring, for a couple of minutes. Gradually whisk in the milk, over the heat, to make a smooth sauce and cook, stirring, for a few minutes until the sauce thickens. Stir in the mustard, a grating of nutmeg and the grated Cheddar until smooth. Take the pan off the heat and season with salt and pepper to taste.

Heat the grill to high. Drain the macaroni as soon as it is ready and add it to the sauce. Stir to combine and then tip into a baking dish.

Sprinkle the Parmesan or other cheese over the surface and place under the grill for about 10 minutes until golden and bubbling. Serve immediately.

Angela's pizza bianco

A classic pizza bianco is one that has no sauce. I have, with the greatest respect, pinched this recipe from my friend Angela Hartnett, whom I have worked with several times over the years. When Angela puts on her 'whites', she immediately looks like a chef. When I put them on, I look like a dentist. So, this *bianco* pizza has a particular resonance for me.

Serves 2 or 4

For the dough

5g active dried yeast
1 tbsp sugar
125ml warm water
250g strong white bread flour
A pinch of sea salt
A drizzle of olive oil

For the topping

200g Taleggio, Suffolk Gold or other semi-soft cheese, sliced
20 salted anchovies
Garlic-and chilli-infused olive oil, or good-quality olive oil
2 eggs
Freshly ground black pepper

For the dough, in a jug, dissolve the yeast and sugar in the warm water and leave until it foams. Mix the flour and salt together in a large bowl and then slowly incorporate the yeast liquid. Add a drizzle of olive oil and mix to a smooth dough. At first the dough will be sticky and difficult to manipulate but persevere and it will come together and form a ball.

Turn the dough on to a clean surface and knead for 5–10 minutes, stretching and folding it vigorously to develop the gluten, as this will give the dough its elasticity. Place in a bowl, cover with a clean tea towel and leave to prove in a warm place for half an hour.

When ready to cook the pizzas, preheat the oven to 220°C/Gas mark 7 and place a couple of baking trays inside to heat up.

Divide the dough into two pieces and roll each out as thinly as you can to a large round. Distribute the cheese and anchovies evenly over each pizza base, then season with pepper and drizzle with garlic-and-chilli oil, or good-quality olive oil if you prefer. Finally, crack an egg into the centre of each pizza. Lightly flour the hot baking trays and carefully lift the pizzas on to them. (Alternatively, you can build the pizzas on floured baking trays and slide them on to the hot trays to cook.)

Bake the pizzas in the oven for 15 minutes. Finish with another drizzle of oil before serving, if you like.

Cheese and potato cake

This is another staple of my Mancunian heritage – more a 'cake' than a 'bake' to me because I have eaten it more often cold, as a snack, than hot as a side dish. But both work extremely well. Sometimes called pan haggerty or simply 'potato bake' in England, a variation of the recipe might be known in France as *truffade*. If you swapped the cheese for 200ml cream, you would have a hybrid version of Dauphinoise or Lyonnaise potatoes.

Serves 4

50g butter
675g floury potatoes, peeled and very thinly sliced
1 medium-large onion, peeled and thinly sliced
200g hard cheese, such as Cornish Gouda, grated
1 tsp freshly grated nutmeg
Sea salt and freshly ground black pepper

Preheat the oven to 190°C/Gas mark 5. Use a small knob of the butter to grease the base of a shallow roasting tin, about 30 x 20cm.

Melt the rest of the butter in a small pan. Put a layer of sliced potatoes in the prepared tin and brush with melted butter. Add a layer of sliced onion, then a layer of grated cheese. Season with salt, pepper and nutmeg. Repeat the layers, finishing with a layer of cheese.

Bake in the oven for 45 minutes or until golden brown on the surface and tender all the way through. Serve hot or allow to cool before eating.

Aligot

Three of my favourite ingredients go to make this stretchy, silky purée of potato, cheese and garlic. A regional French dish from Aubrac in the southern part of the Massif Central, it's not just your average mashed spuds and melted cheese – it has style and panache. Usually it is made with Tomme cheese, from the Auvergne, but any creamy soft cheese works well; I often use Boy Laity, a Cornish Camembert-style cheese. For the mash, use floury potatoes – an interesting variety such as Russet or Yukon Gold if you can get hold of them. Aligot can be served alone or to accompany sausages (it's particularly good with smoked sausages), or seared pig's liver, which is my favourite way to enjoy it.

Serves 4–5

1kg floury potatoes
2 large garlic cloves, peeled and grated
150ml single cream
150g butter
500g semi-soft cheese, such as Boy Laity, sliced
Sea salt and freshly ground black pepper

Peel the potatoes and cut into roughly 5cm pieces. Put them into a large pan, cover with water, add salt and bring to the boil. Cook for about 20 minutes until tender – it should be easy to push a fork into one of the pieces. Drain the potatoes in a colander and leave them to steam-dry for 5 minutes; this will remove enough moisture to ensure they make a light mash.

Tip the potatoes back into the empty pan and use a masher to mash them (or you can use a potato ricer to rice them back into the pan). The mash should still be hot after the 5 minutes of drying but if the temperature has dropped, place the pan over a low heat before adding the other ingredients.

Add the garlic, cream and butter to the potatoes and stir with a wooden spoon to combine. Add the slices of cheese and stir in thoroughly. The mash should become soft and pliable and form lovely long ribbons that stretch from the spoon. Season with salt and pepper to taste and serve straight away.

Lazy dairy bake

Originally this was the filling for a tart. On one occasion when I was trying to cut corners, I made the filling and baked it in a dish without the pastry. The result was very satisfying and it has made many more encores since. The best bit, for me, is that it brings together so many dairy products in one dish. It tastes great, too.

Serves 4–6

25g butter
1 onion, peeled and finely chopped
2 garlic cloves, peeled and finely chopped
200g waxy potatoes, peeled and cut into thin discs
120ml whole milk
120ml crème fraîche
200g beetroot, peeled and grated
125g Cheshire or Caerphilly, grated
125g Cheddar, grated
20g chives, chopped
Sea salt and freshly ground black pepper

Preheat the oven to 200°C/Gas mark 6.

Melt the butter in a pan, add the chopped onion and cook for 10 minutes, until soft and golden brown. Add the garlic and cook for another minute.

Add the sliced potatoes, milk, crème fraîche and some salt and pepper. Simmer for 15–20 minutes until the potatoes are tender. Add the grated beetroot and stir to combine, then remove from the heat. Set aside a handful of each cheese for the topping, then stir the rest into the mix. Check the seasoning.

Spoon the mixture into a medium baking dish and scatter over the reserved cheese and chopped chives. Bake for 10–15 minutes or until the cheese is melted and bubbling with dark, crispy bits around the edges. Serve immediately, with some good, crusty bread if you like.

Ricotta gnocchi

If I was a disc jockey, playing cheesy music on a late-night radio station, I would call myself DJ Ricotta Gnocchi. I might still do it. But for now, this one is for all you cheese-lovers out there. My authentic gnocchi are delicious cooked and served with melted garlic butter or any kind of pasta-appropriate sauce. Of course, you can make them using your own ricotta (see p.102), in which case it's important to let the ricotta drain and dry as much as possible beforehand.

Serves 4

230g ricotta
180g Parmesan or Old Winchester cheese, finely grated
2 medium eggs
120g plain flour
Sea salt and freshly ground black pepper

For the garlic butter
100g butter
1 garlic clove, peeled and grated
A sprig of thyme, leaves stripped (optional)

Put the ricotta, grated cheese, eggs and some salt and pepper into a large bowl and stir until evenly combined. Stir in the flour to form a soft dough.

Divide the dough into 3 or 4 pieces, and roll each piece on a floured surface into a 2cm-thick rope. Cut each rope into 3cm pieces, and place the gnocchi on a lightly floured baking sheet. Place in the fridge until ready to use.

For the garlic butter, melt the butter in a frying pan over a low heat, add the garlic and thyme, if using, and heat gently for a minute or so to infuse; keep warm.

Bring a large pan of salted water to the boil over a medium-high heat. Drop in the gnocchi all at once and simmer for 1–2 minutes until they bob to the surface.

Scoop out the gnocchi with a slotted spoon, drain briefly and add to the garlic butter. Turn the gnocchi to coat in the butter and season with pepper. Divide between warmed dishes and serve at once.

Chilli and cheese muffins

With tangy cheese running through them and riddled with fiery bullets of chilli, these polenta muffins are a family favourite. They are perfect with a steaming bowl of soup, or for a treat on a cold, dark evening in front of the fire.

Makes 6–8

120g plain flour
2 tsp baking powder
½ tsp sea salt
120g dried polenta or cornmeal
½ tsp dried chilli flakes
120g mature hard cheese, such as Junas, grated
1 medium egg, beaten
175ml whole milk
50g butter, melted

Special equipment
12-hole muffin tray

Preheat the oven to 200°C/Gas mark 6. Place 8 paper muffin cases in a muffin tray.

Sift the flour, baking powder and salt together into a large bowl and stir in the polenta. Add the chilli flakes and all but 10g of the cheese. Stir to mix.

Make a well in the middle of the mixture and then pour in the beaten egg, milk and melted butter. Mix together until just combined.

Spoon the mixture into the muffin cases, dividing it evenly. Sprinkle the reserved cheese on top and bake in the oven for 20 minutes until risen and golden. Transfer to a wire rack to cool slightly. Best eaten warm.

Buttermilk cheese scones

This is a great way to use the buttermilk drained off after making butter (see p.35), but you can also buy cultured buttermilk, in which case you won't need to add yoghurt too. The garlicky cream cheese filling adds a delicious savoury edge. If you prefer, you can cut smaller scones to serve as canapés.

Makes 6–8

250g self-raising flour, plus extra for dusting
1 tsp baking powder
40g butter, diced
180g hard cheese, such as Olde Sussex or Old Gloucester, grated
150ml buttermilk
1 tbsp natural yoghurt (optional, see above)
Sea salt and freshly ground black pepper

To serve (optional)
300g cream cheese
¼–½ garlic clove, peeled and finely grated
Cracked black pepper

Preheat the oven to 220°C/Gas mark 7.

Sift the flour, baking powder and a generous pinch of salt into a bowl, then add the butter and rub in with your fingertips until the mixture resembles fine breadcrumbs. (Alternatively, you can do this in a food processor.) Stir in the grated cheese and some pepper.

Add the buttermilk (mixed with the 1 tbsp yoghurt unless you're using cultured buttermilk) and stir until it just comes together and forms a dough.

Transfer the dough to a lightly floured surface, knead lightly and then gently roll out to a 2cm thickness. Using a 5cm plain cutter, stamp out rounds, re-shaping the trimmings to cut a few more. Place on a baking tray and bake in the oven for about 10 minutes until golden and well risen.

Transfer the scones to a wire rack and leave to cool. For the filling, if required, mix the cream cheese with garlic and cracked black pepper to taste.

To serve, split the scones and spread generously with the garlicky cream cheese, or simply with butter if you prefer.

Ricotta and Marsala tarts

These tarts are sweet and melt-in-the-mouth, with a light puff pastry crust. They are inspired by a heavenly tart I had in Italy, made with fresh, local ricotta and a dash of Sicilian Marsala wine. You could, of course, make your own ricotta for the filling (see p.102).

Makes 10–12

For the puff pastry

200g plain flour, plus extra for dusting
A pinch of sea salt
100g cold lard, diced
100g cold butter, diced
2 medium egg yolks
A splash of cold water

For the filling

250g ricotta
1 medium egg, plus 2 egg yolks
40g caster sugar
40ml sweet Marsala wine
Finely grated zest of 1 lemon

Special equipment

12-hole muffin tray

To make the pastry, sift the flour and salt into a bowl, then add the lard and butter and rub in with your fingertips until the mixture resembles fine breadcrumbs. (Alternatively, you can do this in a food processor.) Mix the egg yolks with a splash of water and add to the dry mixture. Mix to combine and bring together to form a stiff dough, adding a little more water if necessary.

Transfer the dough to a lightly floured surface. Roll out to a large rectangle, with a short side facing you. Fold the top third down, then the bottom third up over the top. Give the pastry a quarter-turn then roll and fold again. Repeat this process 4 or 5 times so that the pastry forms a stack of layers. Wrap the dough in cling film and chill in the fridge for 20 minutes.

Preheat the oven to 190°C/Gas mark 5. Roll out the pastry thinly on a lightly floured surface. Stamp out rounds, using a 9cm plain cutter, and use to line the hollows of a muffin tray. Re-roll the trimmings to cut a few more; you should be able to get 10–12 circles.

For the filling, put the ricotta into a large bowl and add the egg, extra yolks, sugar, Marsala and lemon zest. Whisk until the mixture is smooth.

Spoon the filling into the pastry cases and bake for about 20 minutes until risen and golden. Allow to cool slightly in the tray, then carefully ease out the tarts and serve warm, ideally.

Minted lemon mousse

This simple mousse is delicious on its own or served with summer fruits. It also works well as a fancy filling for a brandy snap basket or a light, cheaty cheesecake – with or without a topping of berries. If you have time, make your own fresh curd cheese (see p.95).

Serves 6

50ml cold water
1 tsp powdered gelatine
250g curd cheese
150g natural yoghurt
Finely grated zest of 1 lemon
2 tbsp clear honey
2 sprigs of mint, leaves stripped and finely chopped, plus extra sprigs (optional) to finish
2 medium egg whites

Pour the water into a bowl, sprinkle on the gelatine and leave to soak for 10 minutes until the water is absorbed. Either place the bowl in a microwave on the lowest setting for a minute or two or heat the softened gelatine very gently in a small pan until it has dissolved; do not allow to boil.

In a large bowl, lightly whisk the curd cheese, yoghurt, lemon zest and honey together. Add the dissolved gelatine, whisking as you do so to combine, then stir through the chopped mint. Place in the fridge to chill and thicken for an hour.

Whisk the egg whites in a clean bowl until they form soft peaks, then carefully fold into the chilled mousse, using a spatula or large metal spoon.

Spoon the mousse into small serving glasses or espresso cups and place in the fridge for an hour, or until set. Top with mint leaves, if you like, to serve.

Tiramisu

There are numerous versions of this popular Italian dessert. This one gives a nod to the Southwest, with the inclusion of cider apple brandy. To make it extra special you can, of course, make your own mascarpone (see p.40).

Serves 6–8

2 medium eggs, separated, plus 2 extra yolks
90g caster sugar
5ml vanilla extract
500g mascarpone
20–25 savoiardi (Italian sponge fingers)
250ml freshly brewed, strong black coffee, cooled
60ml cider apple brandy (Kingston Black is my choice)
80g good-quality dark chocolate (80% cocoa solids)

Put the 4 egg yolks, sugar and vanilla extract into a heatproof bowl and set over a pan of simmering water. Whisk until the mixture is pale and thick, then remove the bowl from the pan and continue to whisk until cool.

Once the whisked mixture is cooled, add the mascarpone and whisk briefly until smoothly combined.

In a separate, clean bowl, whisk the egg whites until they form soft peaks, then fold into the mascarpone mixture.

Lay the savoiardi in the bottom of a medium serving dish so they cover the base. Mix the cold coffee and cider apple brandy together and carefully pour over the savoiardi so they soak up the liquid.

Spoon the mascarpone mixture on top of the savoiardi and grate the chocolate evenly over the surface. Place the tiramisu in the fridge to chill for at least an hour before serving.

Baked cheesecake

It's all about the base. Resist the temptation to turn to digestive biscuits for this velvety cheesecake; the coconutty biscuit base is delicious. My advice is to give as much care to the cooling of your cheesecake as you do the cooking – make it a day in advance and leave it to cool overnight. Stop cooking when the filling is still moist, with a little bit of give – it will firm up as it cools.

Serves 6–8

For the base

125g self-raising flour
155g soft light brown sugar
90g desiccated coconut
125g butter, melted, plus extra for greasing

For the filling

500g cream cheese
600ml soured cream
200g caster sugar
3 medium eggs, beaten
Juice of 1 lemon

Special equipment

23cm springform cake tin

Preheat the oven to 180°C/Gas mark 4. To make the base, mix the flour, sugar and desiccated coconut together in a large bowl, then add the melted butter and mix well. Gather the dough and roll out between two sheets of baking parchment to a circle, 1cm thick; remove the top sheet of parchment.

Take the base from the tin and use it as a guide to cut out a circle from the rolled-out dough; don't remove the trimmings. Transfer the parchment with the circular base and the extra dough bits still on it to a baking sheet. Bake for 10 minutes.

Without re-assembling the tin, lightly butter the base and side. Slide the biscuit base on to the cake tin base, re-position the side of the tin and clip together. Keep the extra biscuit bits to one side.

To make the filling, in a large bowl, beat the cream cheese, soured cream and sugar together until smooth, then gradually beat in the eggs and lemon juice. Pour on to the biscuit base in the tin. Use a spatula or palette knife to gently level the surface, making sure there are no gaps between the filling and side of the tin.

Bake for 40 minutes and then switch off the oven, leaving the cheesecake inside. Without opening the door, leave it in the oven for another hour. Remove from the oven, crumble the extra biscuit over the top and allow to cool to room temperature. Refrigerate overnight before unmoulding and cutting into slices to serve.

Useful Things

Troubleshooting guide

Even when you think you have followed everything to the letter in your cheese-making endeavours the outcome is sometimes unexpected. I have experienced this several times while making cheese – occasionally in the form of happy accidents and sometimes as abject failures. Both are valid lessons in the process of turning good-quality milk and cream into dairy derivatives and cheese. Below is a chart listing some common issues and their causes and solutions. Helpful though it is, I hope this section doesn't become the most thumbed part of the book!

PRODUCT	PROBLEM
BUTTER	The cream doesn't separate into butter and buttermilk while it is being churned
	The butter tastes cheesy
YOGHURT	The yoghurt doesn't set after the starter culture is added
	The yoghurt sets but is too thin
	The yoghurt tastes unpleasantly sour

REASON	SOLUTION
The cream hasn't been churned for long enough	Continue to churn
The cream is homogenised, and/or it contains thickeners or stabilisers	Discard and start again, using unhomogenised cream that does not contain thickeners or stabilisers
It was made from cream that was old or out of date	Next time, use fresh cream
The temperature is too low	Adjust the temperature, checking it with a good digital thermometer
The yoghurt starter culture is inactive	It is best to start again if the culture doesn't work. Make sure the culture is active by storing it in the freezer when not in use and checking the best before date. Keep a spare sachet just in case
The milk wasn't hot enough and/or it wasn't heated for long enough	Discard the yoghurt and start again – you could try using a kefir starter this time, if you like
The milk was left for too long after the yoghurt starter culture was added, and/or the temperature was too high	Discard the yoghurt and start again

PRODUCT	PROBLEM
LABNEH	No whey is released
	When preserving labneh balls in olive oil, the balls float rather than submerge
PANEER	The milk doesn't separate into curds
MOZZARELLA	The mozzarella doesn't stretch or melt when you come to eat it or cook with it
FETA	The surface of the feta is sticky or slimy
BRIE-STYLE CHEESE	No white mould appears on the surface of the cheese
	Mould appears on the surface of the cheese, but it is furry
	The rind of the cheese is too thick
WASHED-RIND CHEESE	Unwanted blue mould appears on the rind of the cheese while it is being aged in a 'cave'
	The rind doesn't turn orange

REASON	SOLUTION
You have used a yoghurt that contains cornflour (this is sometimes added as a thickening agent)	Start again, using a natural yoghurt that does not contain cornflour
Too much moisture was left in the labneh when you rolled the balls	Leave the labneh to continue draining for longer
The acidity and/or temperature is too low	Check the temperature with a good digital thermometer, and adjust it if necessary. If you have a pH meter, use this to check the acidity and add more vinegar or lemon juice if you need to
The pH of the milk was too high (i.e. the acidity needed to be increased)	Try adding a touch more starter next time, to increase acidity
Not enough salt was sprinkled on to the curds while they were draining and/or the brine was not salty enough	Next time, take care over salting the curds and ensure that the brine contains enough salt
The temperature is too low	Adjust the temperature
The cheese is too salty	Start again, this time with the correct amount of salt
The cheese is too moist	Wipe the rind and rub on more salt. If you have a hygrometer, confirm that the 'cave' where it is ripening is indeed too moist. Fix this by creating a bit more airflow
Either the variety of *Penicillium candidum* was too strong, or there was too much of it	Next time, reduce the dose of *Penicillium candidum*
The cheese rinds aren't being brine-washed frequently enough, and/or the temperature in the cave is too low	Brine-wash the cheese rinds more often and increase the temperature in the cave slightly
The rind is not getting enough salt because it is not being brine-washed often enough	Brine-wash the cheese rind more often

PRODUCT	PROBLEM
GORGONZOLA-STYLE BLUE CHEESE	Blue veins do not appear in the cheese while it ages
	The cheese tastes bitter
ALL CHEESES WITH RENNET	The curds have set but when you come to cut them, they don't break cleanly
	The curds break into small pieces
HARD CHEESE	The curds don't knit together when they are being heated (after cutting)
	The cheese is developing a dry or cracked rind while it ages in the 'cave'
	The rind slips (falls away from the paste)

REASON	SOLUTION
The cheese doesn't have enough piercings	Pierce the cheese more
The cheese was pressed too hard into the mould	Next time, be gentle when putting the curds into the mould
Too much starter culture and/or rennet was added to the milk at the beginning	Next time, don't use so much of your starter culture or rennet
The starter culture or the rennet was inactive (possibly because it was old)	Start again. Take care to use only high-quality ingredients that are within their sell-by date
The starter culture or the rennet was not stirred into the milk properly	Stir in a little more rennet
The milk was not high-quality	Start again. Use fresh milk, either whole or raw
The curds were heated too quickly	Start again, and this time monitor the temperature carefully
The milk that the curds were made from was too old	Start again with fresher milk
The humidity in the cave is too low	Cover the cheese in an upturned plastic container to create a more humid environment
When the starter was added, acidification happened too fast	Start again or see out the result – your cheese may just have a sour taste
The curds were drained too slowly	Next time, use a mould with larger drainage holes. You can finish the cheese off, it might simply have a sour taste

Glossary

As with all artisan food production, cheese-making comes with its own terminology. Here is a guide:

Acidity The ratio of lactic acid present in fermenting milk or whey. The level of acidity can be read by a pH meter (as acidity goes up, the pH goes down).

Affinage The handling and care of cheese as it ages. Professional *affineurs* buy young cheeses and age them in their own maturation cellars.

Ageing; also ripening Maturing cheese in a cave until it has acquired the perfect texture and fullest flavour.

Bacteria Living organisms that are present in milk. 'Good' bacteria are essential for the development of flavour and texture in dairy products.

Bandage Strips of loose-weave cotton gauze that are used for wrapping around new cheeses.

Bloom The white, fluffy rind on soft cheeses, such as Camembert and Brie, consisting of a complex community of moulds and/or yeasts.

Blown A cheese is 'blown' when the paste has holes in it caused by bad bacteria. The bacteria produce an excess of hydrogen gas and butyric or acetic acid, which causes structural damage to the paste and gives off an unpleasant smell, reminiscent of vomit. This renders the cheese inedible.

Blue A cheese is 'blue' if it features a mould such as *Penicillium roqueforti*. This takes the form of either 'veins' running through the paste or a bloom on the rind.

Brining Cheeses can be rubbed or immersed in a brine, which is usually made by adding salt to whey or water.

Butter The solid part of coagulated cream.

Butterfat The fat contained in whole milk. Usually 2.5–5.5% of the milk is butterfat.

Buttermilk The liquid that's left over when cream is made into butter.

Casein A type of protein contained in milk. The casein proteins in a batch of milk will coagulate over time – a process that is accelerated during cheese-making.

Cheese iron A probe that can be used to draw out a sample from a large cheese. This is generally used to assess hard cheeses as they age.

Churn A metal container for storing milk in dairies.

Coagulation; also curdling When milk splits and solid curds start to separate from the liquid whey.

In cheese-making, coagulation is triggered by adding an acidic starter culture (and usually some rennet) to a batch of heated milk.

Curd The solid that is formed when milk coagulates; it is made up of protein and fat. The curd can be moulded and turned into cheese.

Curdling *see* Coagulation

Cutting The stage at which a freshly formed curd is cut into smaller pieces.

Drum A cylinder-shaped cheese.

DVI (Direct Vat Inoculation) This is a commercial freeze-dried type of starter culture, often favoured for its ease of use.

Enzyme A type of protein that triggers change. It is the enzymes in rennet that help milk to coagulate. Enzymes can also play a role in the ripening of a cheese, and in the development of mould on its rind.

Fat Milk is made up primarily of fat and protein: these are the two key components of cheese and other dairy products.

Fermentation This is the stage at which an ingredient starts to break down. Cheese, yoghurt, soured cream and crème fraîche are all the result of milk fermenting. In cheese-making, inoculation starts the fermentation process, which is the stage when coagulation occurs.

Flocculation This is the moment when the curds start to form in coagulating milk.

Follower A wooden or plastic slab (often disc-shaped) that is placed between a cheese and a press during the 'moulding' stage.

Form *see* Mould (2)

Homogenisation The commercial treatment of milk to distribute its fat globules evenly throughout the liquid. This stops the fat separating from the milk and renders cheese-making impossible.

Inoculation The introduction of acid or a starter culture (and rennet) into heated milk at the start of the cheese-making process. (*See also* Fermentation.)

Ironing Extracting a sample from an ageing cheese to check on progress. (*See also* Cheese iron.)

Lactic acid When an acid or starter culture is added to milk, lactose is converted into lactic acid as part of the fermenting process. Lactic acid helps to protect a cheese from unwelcome bacteria.

Lactose The type of sugar that is naturally found in milk.

Listeria A type of bad bacteria that can develop in raw milk if it is not properly handled.

Maturation *see* Ageing

Mesophile A starter culture that will thrive in warm, but not hot, conditions. (*See also* Thermophile.)

Milling The process of shredding curd before moulding it (used for making Cheddar).

Mould (1) Microscopic fungi that are present everywhere in the atmosphere. They form the 'bloom' on the surface or 'veins' running through a cheese.

Geotrichum candidum Somewhere between a yeast and a mould, this is used to develop flavour and texture in soft-ripened cheeses, such as Camembert

Penicillium candidum This mould creates a bloomy white coating on cheese rinds

Penicillium roqueforti The most common 'blue' mould

Mould (2) Also known as a 'hoop' or 'form', a mould is used to shape curds in cheese-making. Excess whey is drained off during moulding.

Mucor A type of black furry mould that can develop on a cheese, spoiling its appearance. On a hard rind-aged cheese this can be washed or wiped off and the cheese can still be eaten, but a soft cheese with mucor should be discarded.

Needling Piercing the body of a cheese, to let air enter and encourage the growth of blue mould. This leads to the development of the 'veins' in a blue cheese.

Paste The inside of a cheese (as opposed to the rind).

Pasteurisation Heating milk to a certain temperature in order to destroy potentially harmful bacteria.

Pathogens Bacteria or other micro-organisms that can cause disease.

PDO (Protected Designation of Origin) Protects the name of a product, with regard to where it was made. So in order to call a cheese 'Roquefort', it must be made in the town of Roquefort in France.

Peg mill The traditional machine that shreds curd during Cheddar-making.

PGI (Protected Geographical Indication) Protects the identity of a product, with regard to the area where it is produced. Dorset Blue Vinny, for example, can only be labelled as such if it is made in Dorset.

Phage A virus that can attack the bacteria in milk if a particular strain of culture loses its strength through over-use. Cheese producers often swap small amounts of culture to keep the strain vibrant and alive, so it is less likely to be affected. Phage incidents are very rare but do present a real risk.

Pitch The moment when a cheese-maker stops stirring the curd and lets it settle at the bottom of the vat.

Pressing Using a cheese press to compact cheese and push out residual whey during 'moulding'.

Raw milk Milk that has not been heat-treated.

Rennet The enzyme used to inoculate milk, usually after a starter culture has been added. Rennet traditionally comes from the stomach of a young cow but it can also be produced using vegetarian ingredients.

Rind The outer skin of a cheese.

Ripening *see* Ageing

Scalding The heating of the curd during cheese-making.

Starter culture An acidic bacteria culture that is added to heated milk in order to kick-start coagulation.

Territorial A generic term for traditional cheeses from a specific area, such as Lancashire, Cheshire or Gloucester.

Terroir The natural environment in which a cheese is created, including the pasture the animals graze on to produce the milk.

Thermophile A heat-tolerant starter culture. (*See also* Mesophile.)

Truckle A small cylinder of cheese, normally weighing 1.5 kilos.

Vat A container used for making cheese.

Washing Rubbing the rind of a cheese with a liquid, usually brine.

Wheel A large round cheese, such as Brie, which is usually flat.

Whey The liquid by-product that is formed when milk coagulates after inoculation.

Directory

There is a host of good information on making cheese and dairy products out there, both online and in book form. Whether it is practical help, equipment or recipes that you need, there are many websites with forums that will be of use to the budding cheese-maker. The sources that I most frequently turn to are:

Useful websites

cheesemaking.co.uk
The Moorland Cheesemakers Ltd website offers practical advice and recipes. It has an online shop selling all the kit needed to create fantastic products.

cheesemakingshop.co.uk
You can buy a whole range of cheese-making ingredients and equipment from The Cheese Making Shop, including rennet and starters.

rawmilk.simkin.co.uk
You can use this online directory to find a raw milk supplier near you.

nealsyarddairy.co.uk
Here you can buy a huge range of artisan British cheeses, as well as finding out in-depth information about them.

britishcheeseawards.com
The British Cheese Awards celebrate the best of cheese-making in the UK. This website lists current winners and details about future events.

Further reading

The Art of Natural Cheesemaking
by David Asher
published by Chelsea Green Publishing

Home-made Cheese, Artisan Cheesemaking Made Simple
by Paul Thomas
published by Lorenz Books

Acknowledgements

I love the solitary aspect of writing, the bit where you sit quietly in front of a blank computer screen, looking to turn the information in your head into written words – occasionally looking for inspiration by staring out of the window, drinking a freshly brewed coffee or practising acceptance speeches in the mirror for awards that will never come. However, it is the process that follows which really makes a book. Cheese has always been a major pleasure in my life but it is only with the encouragement of some talented and patient people that I have been able to turn a great interest into a book. I would like to thank the following people:

Gavin Kingcombe is a photographer and filmmaker. He is such a pleasure to work with and it was great continuing our creative partnership. This is as much his book as it is mine.

Sam Lomas is a very skilled chef and teacher. Without Sam's input this book wouldn't be the same.

Special thanks to the Kitchen Team at RCHQ: Gelf Alderson, Andy Tyrrell, Connor Reed, Mark McCabe, Ben Harrison, Charlie Gabriel, Bob Hains, Marcus Campbell and Jamie Butler.

At Bloomsbury, my editor Xa Shaw Stewart has led this project brilliantly and kept me on track.

Project editor Janet Illsley and designer Will Webb combine wonderfully on detail and design. Nikki Duffy takes my words and polishes them into shape; she's so good at that. I'd especially like to thank Hugh Fearnley-Whittingstall for his inspiration, and Antony Topping of Greene and Heaton.

Thanks also to: Murry Toms, for all the best cheese club evenings and for being my great friend; Wendy and Wayne for Camp Compton, where this book took shape; Adam Raimes for the aligot inspiration; Mark Hix for cheese sauce tips; and Francis Nesbitt for his energy, humour and friendship.

And to my parents Jean and Brian, as well as my brother Ian and his family, who still don't really know what I do.

Lastly, my beautiful wife Elli who is the centre of my world, and our three magic girls, Agnes, Betsy and Jeanie-Ray.

Index

Page numbers in *italic* refer to the illustrations